Jing Wu
I-Shin Chang

中国环境管理　政策与制度

Environmental Management in China

Policies and Institutions

 Chemical Industry Press

内容简介

《中国环境管理：政策与制度》（*Environmental Management in China：Policies and Institutions*）以英文形式全面系统总结了中国当代环境管理政策与制度的发展历程，生动展现真实、立体的中国环境管理全貌，为推动一带一路绿色发展传播中国声音。全书共19章，第1章是环境管理行政管理的演化；第2章阐述了环境管理指导思想的产生和发展；第3章至第16章介绍我国主要环境管理政策与制度的发展历程、特点及其成效，包括环境规划制度、环境影响评价制度、三同时制度、排污收费和环境保护税制度、环境保护目标责任制、排污许可制度、污染集中控制、总量控制制度、公众参与制度与环境信息公开制度、生态保护红线制度、区域联防联控、生态环境损害赔偿制度、清洁生产制度、生态补偿制度等；从第17章到第19章，介绍了我国环境保护在大气、水和土壤方面的重要任务。最后给出了主要法律、法规、条例、措施等的中英文对照表，以供参考。

本书可供高等院校环境专业作为教材使用，还可供从事环境保护、环境管理等领域的科研人员阅读参考。

本书由化学工业出版社与 Springer Nature 出版公司合作出版。本版本仅限在中国内地（大陆）销售，不得销往中国台湾地区和中国香港、澳门特别行政区。

图书在版编目（CIP）数据

中国环境管理：政策与制度 =Environmental Management in China：Policies and Institutions：英文 / 吴婧，张一心著 . — 北京：化学工业出版社，2020.9
ISBN 978-7-122-37588-9

Ⅰ．①中… Ⅱ．①吴… ②张… Ⅲ．①环境管理 - 研究 - 中国 - 英文 Ⅳ．① X321.2

中国版本图书馆 CIP 数据核字（2020）第 158775 号

责任编辑：满悦芝	文字编辑：王 琪
责任校对：宋 玮	装帧设计：张 辉

出版发行：化学工业出版社
（北京市东城区青年湖南街13号　邮政编码100011）
印　　装：涿州市般润文化传播有限公司
710mm×1000mm　1/16　印张18½　字数403千字　2020年10月北京第1版第1次印刷

购书咨询：010-64518888　　　　　售后服务：010-64518899
网　　址：http://www.cip.com.cn
凡购买本书，如有缺损质量问题，本社销售中心负责调换。

定　　价：188.00元　　　　　　　　　　　　　　　版权所有　违者必究

Jing Wu · I-Shin Chang

Environmental Management in China

Policies and Institutions

🔷 Chemical Industry Press

Foreword

This book was written for college students both at the graduate and undergraduate level, as well for anyone interested in learning about China's environmental management. The core of this book is to comprehensively and systematically summarize the policies and systems development of China's environmental management. It has been more than 40 years since the initiation of China's environmental management through participating in the United Nations Conference on the Human Environment (the Stockholm Conference) in 1972. With reform and opening up, fast national economic growth, prompt social progress, and increasing public environmental awareness, the guiding ideology of China's environmental management has undergone the evolution from the basic national policy to the sustainable development strategy, to the scientific outlook on development, and to the ecological civilization. Meanwhile, through continuous evolution, retrospection, and innovation, the environmental policies and the environmental management system are becoming more and more matured, gradually.

In order to promote the socialism with Chinese characteristics and to build a sustainable and beautiful China, the "Five-in-One" Strategy was proposed during the 18th National Congress of the Communist Party of China (the NCCPC), in which ecological civilization construction, economic construction, political construction, cultural construction, and social construction shall be thoroughly integrated, to further facilitate and expedite the development of China's environmental management, with the more explicit goal, more thorough system, and more rigorous content. In general, since the 18th NCCPC, the core thoughts of China's environmental management are to improve the environmental quality, to facilitate collaborative environmental governance based on the most stringent institutional construction, to strengthen the liability system of environmental protection, and to promote development transformation driven by environmental protection, so as to achieve the green development goals.

This book has presented the authentic, multi-faceted, and overall perspective of China's environmental management and demonstrated China's determination and efforts in promoting green development of Belt and Road Initiative. Greatly benefited from the precious experiences of Western Countries and the practical experiences of exploration and implementation, domestically, through these years, China's environmental management has gradually developed into a feasible, reliable, and effective system. Especially, based on the idea of sustainable development, the ecological civilization construction was proposed as China's Scheme to properly manage the relationship between environment and development, so as to better build Chinese Spirit, to reflect Chinese Value, to demonstrate Chinese Strength, and to realize Chinese Dream.

There are 19 chapters in this book. Chapter 1 is to introduce the evolution and transformation of the administration of environmental management. Chapter 2 is to illustrate the initiation and development of the guiding ideology of environmental management. From Chaps. 3 to 16 are to summarize the introduction, the development, the characteristics, the features and the effectiveness of key policies and systems of environmental management. From Chaps. 17 to 19 are to demonstrate the crucial tasks of environmental protection in atmosphere, water, and soil. At last, a Comparison Table of key laws, regulations, ordinances, measures, and so forth (English to Chinese) is appended to provide a quick reference.

Preface

In China, environmental protection work was formally commenced in 1973. Through continuous learning process of exploration, execution, review, and amendment on the implementation of environmental protection work, a set of environmental management system, conforming to China's national conditions, has been gradually established to provide effective assurance for strengthening environmental protection. During the initial stage of China's environmental protection work, from 1973 to 1979, the "Three Synchronizations", the "Environmental Impact Assessment", and the "Emission Fees for Noncompliance", also known as the "Three Old Management Schemes", were put forward and implemented to find the basis of administrative management system for China's environmental protection work and to denote the institutionalization of China's environmental protection work. In addition, the Three Old Management Schemes had generated significant effects on pollution prevention and control for existing and new pollution sources to greatly promote the establishment and execution of China's environmental protection work. In 1989, during the Third National Conference on Environmental Protection, through the experience learned from the practice of environmental protection work and the establishment of environmental management system, the idea that the pathway of China's environmental protection work should be in line with China's national conditions was put forward. Thus, five different environmental management schemes, including the "Liability System for Environmental Protection Objectives", the "Quantitative Performance Evaluation on Comprehensive Environmental Governance for Urban Environment", the "Centralization of Pollution Control", "Deadline for Attainment", and the "Emission Permit", also known as the "Five New Management Schemes", were promulgated. And the basic institutional framework of China's environmental management was then founded on these eight environmental management schemes, including the Three Old Management Schemes and the Five New Management Schemes.

In the twenty-first century, as the emphasis of environmental protection was gradually shifted from governing pollution to improving environmental quality and promoting environmental service functions, the orientation of environmental management was inevitably adjusted from pollution control to the enhancement of environmental quality. And, the perception of environmental protection was then changed from abating pollution to preventing pollution, including: To put environmental protection first, to promote environmental service functions, and to enhance the production capacity of ecological goods. Consequently, several effective environmental management systems and measures were progressively promulgated and implemented. For example, in the "Total Emission Cap", emission cap was determined for major pollutants as the binding indicator to be incorporated into the National Economic and Social Development Plan. In the "Environmental Monitoring and Emergency Response Management System", six regional Environmental Protection Supervision Centers, including North, East, South, North East, North West, and South West, were set up, directly under the Ministry of Environmental Protection (the MEP), to deal with the cross-regional and cross-watershed environmental regulatory issues. In the *Law of the People's Republic of China on Promoting Cleaner Production* and the *Law of the People's Republic of China on Promoting Circular Economy*, pollution control on sources and processes was stipulated. In the "Information Disclosure and Public Participation", environmental information disclosure and public participation in environmental impact assessment were continuously strengthened. In the "Environmental Economy Policy System", a framework consisting of green credit, insurance, trade, price of electricity, stock, taxation, etc. was constructed prelimi-narily. Meanwhile, some environmental management schemes incompatible with social and economic development were gradually phased out. For example, "Quantitative Performance Evaluation on Integrated Environmental Governance for Cities" was terminated in 2012, and the "Deadline for Attainment" was deleted from the 2014 Amendment of the *Law of the People's Republic of China on Environmental Protection* (the EP Law 2014).

Tianjin, China Jing Wu
Huhhot, China I-Shin Chang

Acknowledgments

First of all, with hearty gratitude, the authors would like to acknowledge the individuals and institutions that have made this book possible. Many thanks to the financial support from the Key Funding Projects of the Silk Road Literary Program and the Institute of Ecological Civilization of Nankai University. In addition, Ms. Yilin Chen, Ms. Wenqi Wang, Mr. Tong Li, Ms. Chonglin Ran, Ms. Mengwei Zeng, and Ms. Qian Li made great contributions to data collecting, processing and analyzing, and tabulating and drawing. Finally, the authors would like to express their most sincere appreciation toward the Executive Editor, Ms. Yuezhi Man, at Chemical Industry Press and the Executive Editor at Springer, for their tremendous support to publish this book.

Contents

Chapter 1 Administration on Environmental Management 1
  1.1   The History of the MEE 1
  1.2   Structure and Functions of the MEE 3

Chapter 2 Evolution of China's Environmental Policy 7
  2.1   In the 1970s: The Principles of National Environmental
       Protection 7
  2.2   In the 1980s: The Principles of Three-Simultaneity
       and Three-Unification 8
  2.3   In the 1990s: Sustainable Development 9
  2.4   In the 2000s: Scientific Outlook on Development 10
  2.5   In the 2010s: Ecological Civilization 12
  2.6   Concluding Remarks 14
  Reference 15

Chapter 3 Environmental Planning 17
  3.1   The Development of Environmental Planning 17
  3.2   Environmental Protection and the Plan of National
       Economic and Social Development 21
       3.2.1   Environmental Protection During the 6th FYP
              (1981–1985) 23
       3.2.2   Environmental Protection During the 7th FYP
              (1986–1990) 24
       3.2.3   Environmental Protection During the 8th FYP
              (1991–1995) 25
       3.2.4   Environmental Protection During the 9th FYP
              (1996–2000) 25
       3.2.5   Environmental Protection During the 10th FYP
              (2001–2005) 26
       3.2.6   Environmental Protection During the 11th FYP
              (2006–2010) 28

		3.2.7	Environmental Protection During the 12th FYP (2011–2015)	29
		3.2.8	Ecological and Environmental Protection During the 13th FYP (2016–2020)	30
	3.3	The Coordination and Integration of Multiple Plans		31
	References			34

Chapter 4		**Environmental Impact Assessment**		35
	4.1	The Concept and Principles of EIA		35
	4.2	The Development of EIA Globally		36
	4.3	The Development of EIA in China		37
		4.3.1	Initial Phase (1973–1978)	37
		4.3.2	Early Implementation Phase (1979–1989)	37
		4.3.3	Improvement Phase (1990–2002)	39
		4.3.4	Breakthrough Phase (2003–2014)	41
		4.3.5	Reform (2015–Date)	42
	4.4	The Legal Framework of EIA in China		43
	4.5	EIA for Project (Project EIA)		45
		4.5.1	Categorization for Project EIA	46
		4.5.2	Process for Project EIA	46
		4.5.3	Documentation for Project EIA	46
		4.5.4	Public Participation in Project EIA	48
		4.5.5	Review and Decision on the EIA Documents	49
		4.5.6	EIA Follow-up for Construction Project	52
	4.6	Regional EIA (REIA)		53
	4.7	EIA for Plan (PEIA)		54
		4.7.1	Applicable Scope of PEIA	54
		4.7.2	Roles and Responsibilities	55
		4.7.3	Approaches and Requirements of PEIA	56
		4.7.4	Public Participation in PEIA	56
		4.7.5	Submission and Review of the PEIA Documents	58
		4.7.6	Approval of Plan	58
		4.7.7	EIA Follow-up for Plan	58
	4.8	Certificate System of EIA Consultancy (Revoked in 2018)		59
	4.9	Management of Qualified EIA Practitioners		60
	4.10	Prospects of EIA in China		61
	References			62

Chapter 5		**Three Synchronizations System**		63
	5.1	The Development of the Three Synchronizations System		63
	5.2	The Requirements of the Three Synchronizations System		66
		5.2.1	Synchronous Design	66
		5.2.2	Synchronous Construction	67
		5.2.3	Synchronous Operation	67

		5.2.4	Roles and Responsibilities	67
		5.2.5	The Violation	68
	References			68
Chapter 6	**Emissions Charges System and Environmental Protection Tax System**			**69**
	6.1	Development of the Emissions Charges System		69
	6.2	The Content of the Emissions Charges System		72
		6.2.1	The Payer	72
		6.2.2	Category and Amount of Pollutant Emission	72
		6.2.3	Category of the Emissions Charges	72
		6.2.4	Charge Rates	73
		6.2.5	Collection and Utilizations of the Emissions Charges	73
	6.3	The Environmental Protection Tax System		75
		6.3.1	Who Pays the Environmental Protection Tax	80
		6.3.2	Items Subject to the Environmental Protection Tax	80
		6.3.3	Basis and Rate of the Environmental Protection Tax	82
		6.3.4	Preferential Policy	83
		6.3.5	Monitoring and Enforcement	83
	References			84
Chapter 7	**Target Responsibility System of Environmental Protection and Performance Evaluation System**			**85**
	7.1	The Formation and Development of the Target Responsibility System of Environmental Protection		85
	7.2	Targets of Environmental Protection and Indicators Setting		88
	7.3	Target Setting		88
	7.4	Performance Evaluation		88
	7.5	Shared Responsibility of Ecological and Environmental Protection for Government and Party		90
	7.6	The Audit of Natural Resources Assets for Leading Cadres While Leaving Office		92
	References			93
Chapter 8	**Centralized Pollution Control System**			**95**
	8.1	Centralized Heating in Urban Areas		96
	8.2	Centralized Pollution Control of Wastewater		97
	8.3	Centralized Pollution Control of Municipal Solid Waste		100
		8.3.1	Landfill	101
		8.3.2	Incineration	101
	References			103

Chapter 9 Emission Reporting, Registration and Permit System 105
- 9.1 The Formation and Development of the Emission Permit System 105
- 9.2 Implementation Process of the Emission Reporting, Registration, and Permit System 108
  - 9.2.1 Reporting and Registering Pollutant Emission 108
  - 9.2.2 Verifying Pollutant Emission.................. 108
  - 9.2.3 Issuing the Emission Permit 108
  - 9.2.4 Supervision and Management for the Emission Permit 109
- 9.3 Essentials of the Emission Permit System.............. 109
  - 9.3.1 Who Shall Apply for the Emission Permit 109
  - 9.3.2 An Integrated Emission Permit 109
  - 9.3.3 Procedure to Apply for the Emission Permit 109
  - 9.3.4 Conditions for Obtaining an Emission Permit........ 110
  - 9.3.5 Evaluate and Determine the Allowable Emission Based on EIA Results...................... 110
  - 9.3.6 Self-reporting and Self-monitoring.............. 110
  - 9.3.7 Information Disclosure and Public Supervision 111
  - 9.3.8 Enforcement and Responsibilities 111
- 9.4 Pilot Program of Emissions Trading System.............. 113
- References 114

Chapter 10 Emissions Cap System 115
- 10.1 Establishment and Development of the Emissions Cap System 115
- 10.2 Objective Classifications of Emissions Cap............... 119
  - 10.2.1 Emissions Cap Based on Environmental Capacity 119
  - 10.2.2 Emissions Cap Based on Environmental Target 119
  - 10.2.3 Emissions Cap Based on Previous Emissions........ 120
  - 10.2.4 Emissions Cap Based on Cost Benefit Analysis 120
- 10.3 Pollutants Classification of Emissions Cap 120
  - 10.3.1 Emissions Cap of Water Pollutants 121
  - 10.3.2 Emissions Cap of Air Pollutants 121
  - 10.3.3 Emissions Cap of Pollutants Discharged into the Sea 122
- 10.4 Measures of Emissions Cap........................ 122
  - 10.4.1 Evaluations on the Emissions Cap System During the 11th FYP 122
  - 10.4.2 Target Setting 124
  - 10.4.3 Verification and Auditing 124
- 10.5 Final Remarks 125
- References 126

Contents

Chapter 11 Joint Pollution Prevention and Control 127
- 11.1 The Need for Joint Air Pollution Prevention and Control 127
- 11.2 Key Regions and Key Points of Joint Air Pollution Prevention and Control 129
- 11.3 Key Tasks of Joint Air Pollution Prevention and Control 129
  - 11.3.1 Optimizing the Regional Industrial Structure and Layout 129
  - 11.3.2 Intensifying the Extent of Pollution Prevention and Control on Key Pollutants 130
  - 11.3.3 Promoting the Comprehensive Utilization of Clean Energy 131
  - 11.3.4 Intensifying Pollution Prevention and Control of Motor Vehicles 133
  - 11.3.5 Improving the Regional Air Quality Monitoring System 134
- References 135

Chapter 12 Public Participation and Environmental Information Disclosure 137
- 12.1 Legislative Process of Public Participation in Environmental Management 137
- 12.2 Channels of Public Participation in Environmental Management 141
- 12.3 Other Means of Public Participation in Environmental Management 143
- 12.4 Environmental Information Disclosure 146
- References 149

Chapter 13 Restraining Redlines System of Ecological and Environmental Protection 151
- 13.1 The Initiation and Development 151
- 13.2 Practices of Redline in Resources and Environment 153
  - 13.2.1 Cultivated Land Redline 153
  - 13.2.2 Water Resource Redline 154
  - 13.2.3 Emission Cap Redline 155
- 13.3 Framework Design for Redline in Resources and Environment 155
  - 13.3.1 Upper Limit of Resource Consumption (Resource Consumption Cap) 156
  - 13.3.2 Bottom Line of Environmental Quality (Environmental Quality Baseline) 157
  - 13.3.3 Ecological Protection Redline 158
- References 158

Chapter 14 Compensation for Environmental Damage 161
14.1 International Experience 162
  14.1.1 In the United States 162
  14.1.2 In the European Union 163
14.2 Necessity 163
14.3 Related Policies 164
  14.3.1 Scope of Application 165
  14.3.2 Scope of Compensation 165
  14.3.3 Claimant to Compensation 165
References 166

Chapter 15 Cleaner Production 167
15.1 Initiation 168
15.2 Legal System Development 168
  15.2.1 The Beginning Stage (1973–1992) 168
  15.2.2 The Legalization Stage (1993–2002) 169
  15.2.3 The Institutionalization Stage (2003–2005) ... 170
  15.2.4 The Perfection Stage (2006 to Date) 170
15.3 Capacity Building 171
  15.3.1 Institution Development 171
  15.3.2 Professional Training 174
  15.3.3 Guidelines and Standards 175
15.4 Mandatory Cleaner Production Audit 175
15.5 Recommendations for Promoting CP Development
     in China 177
15.6 Conclusions 179
References 179

Chapter 16 Ecological Compensation 181
16.1 Introduction 181
16.2 Development of ECMs 182
16.3 Financial Sources for ECM in China 184
  16.3.1 Financial Transfer Payment 184
  16.3.2 Special Funds 186
  16.3.3 Resource Taxation System 186
  16.3.4 Payments for Ecological Compensation 187
16.4 Major Ecological Engineering Projects in China 188
  16.4.1 Three-North Project 188
  16.4.2 Natural Forest Protection 189
  16.4.3 Beijing-Tianjin Sandstorm Sources Governance . 189
  16.4.4 Returning Farmland to Forest/Grassland 189
  16.4.5 Returning Rangeland to Forest/Grassland 190

	16.5	ECM for Sector	190
		16.5.1 ECM for Forest	190
		16.5.2 ECM for Coal Mining	194
		16.5.3 ECM for Grassland	197
	16.6	ECM at Local Level	198
		16.6.1 ECM for the Three Rivers Source District, Qinghai Province	198
		16.6.2 ECM for Poyang Lake, Jiangxi Province	199
		16.6.3 ECM for Nature Reserves, Hainan Province	199
		16.6.4 The ECM for Natural Resource Utilization, Sichuan Province	200
		16.6.5 The ECM for Watershed Conservation, Yunnan Province	200
	16.7	Conclusions	202
	References		202
Chapter 17	**Air Quality Governance in China**		**205**
	17.1	General Features of Air Pollution	205
	17.2	Legal Institution and System	206
	17.3	Process of Development	209
		17.3.1 Phase I: Legal Institution Management for Air Pollution Control on Soot (from 1979 to 1997)	210
		17.3.2 Phase II: Exploring Unified Regional Air Pollution Prevention and Control for Emerging Compound Pollution (from 1998 to 2012)	212
		17.3.3 Phase III: Integrating Unified Regional Air Pollution Prevention and Control for Compound Pollution (from 2013 to Date)	214
	17.4	Evaluations and Prospects	217
	References		218
Chapter 18	**Water Quality Governance in China**		**219**
	18.1	Water Resources in China	219
	18.2	Water Pollution in China	221
	18.3	Laws and Regulations on Water Pollution Prevention and Control	223
	18.4	Water Management in China	225
	18.5	Progress	229
	References		231
Chapter 19	**Land Resources Governance in China**		**233**
	19.1	Land Resources in China	233
	19.2	Land Resources Protection	235

	19.3	Land Resources Degradation and Land Pollution in China	236
		19.3.1 Land Degradation	236
		19.3.2 Land Pollution	236
		19.3.3 Overall Spatial Distribution of Soil Pollution	237
		19.3.4 Status Quo of Soil Pollution of Various Types of Land	237
	19.4	Laws and Regulations on Land Pollution Prevention and Control	238
	19.5	Challenges and Countermeasures	244
		19.5.1 Challenges	244
		19.5.2 Countermeasures	245
References			249
Comparison Tables			251

List of Figures

Fig. 1.1	Structure of the ministry of ecology and environment	4
Fig. 2.1	The development process of China's environmental policy (1973–2018)	15
Fig. 3.1	The scope of plans for coordination and integration	34
Fig. 4.1	The process for project EIA. *Source* Technical guideline for environmental impact assessment of construction project—general program (HJ 2.1—2016)	47
Fig. 4.2	Evolution of public participation and information disclosure in EIA in China	49
Fig. 4.3	Number of certified EIA consultancies (2008–2018)	60
Fig. 5.1	Different timing for various environmental management systems	66
Fig. 6.1	Emissions charges collected from 1979–2015	75
Fig. 6.2	Evolving process from emissions charges to environmental protection tax in China	77
Fig. 8.1	Areas and amount of heating supply of centralized heating system during winter time in China. *Data Source* The Ministry of Housing and Urban-Rural Development of the People's Republic of China, *the Statistical Yearbook of Urban and Rural Development.* and *the Bulletin of Urban and Rural Construction Statistics*	96
Fig. 8.2	Wastewater discharge in China from 2000 to 2015. *Data Source* Data on processing capacity of WWTP were extracted from the National Bureau of Statistics. The number of wastewater treatment plant (WWTP) was acquired from *the Statistical Bulletin of Urban and Rural Construction* issued by the Ministry of Housing and Urban-rural Development of the People's Republic of China.	98

Fig. 8.3	Wastewater treatment facilities and capacity in China from 2000 to 2017. *Data Source* Processing capacity of WWTPs, Nation Bureau of Statistics. The number of wastewater treatment plant (WWTP): from *the Bulletin of Urban and Rural Construction Statistics* issued by the Ministry of Housing and Urban-rural Development of the People's Republic of China.	99
Fig. 8.4	Municipal solid waste generation and urban population in China from 1980–2016. *Data Source* Nation Bureau of Statistics.	100
Fig. 10.1	Compiling process of the plan for the emissions cap of major pollutants during the 12th FYP	123
Fig. 10.2	Emission cap of SO_2 for each province during the 12th FYP *Note* The national emissions cap of SO_2 by 2015 was 2.0864×10^7 tons, in which 2.0674×10^7 tons was allocated to provinces and 1.90×10^5 tons was reserved for the pilot program of emissions trading of SO_2	124
Fig. 10.3	Emission cap of COD for each province during the 12th FYP *Note* The national emissions cap of COD by 2015 was 2.3476×10^7 tons (including 1.2219×10^7 tons for industrial and domestic sources), in which 2.3352×10^7 tons was allocated to provinces (including 1.2146×10^7 tons for industrial and domestic sources) and 1.24×10^5 tons was reserved for the pilot program of emissions trading of COD.	125
Fig. 13.1	Composition of ecological redline system.	156
Fig. 15.1	Number of CP centers and CP service institutions, and persons to complete CPA auditor training	174
Fig. 15.2	Number of companies to implement MCPA and complete CPA assessment and acceptance.	176

List of Tables

Table 1.1	The evolutionary process of environmental protection administration in the central government	3
Table 1.2	Affiliated public institutions and social groups directly led by the MEE	4
Table 2.1	The targets of ecological civilization by 2020	13
Table 3.1	Categories of environmental protection plan	21
Table 3.2	Contents related to environmental planning in selected environmental law and regulations	22
Table 3.3	Major targets of ecological and environmental protection during the 13th FYP	32
Table 4.1	List of various EIA guidelines in China (updated in January 2019)	40
Table 4.2	Legal framework on EIA in China (updated in January 2019)	44
Table 4.3	Key EIA stages and the requirements of public participation	50
Table 4.4	EIA requirements for integrated plans and specific plans	57
Table 6.1	Changes of emissions charges standard for SO_2	74
Table 6.2	Emissions charges standards for different pollutants promulgated by various provinces/cities by 2016	75
Table 6.3	Comparison of the emissions charges system and the environmental protection tax system	78
Table 6.4	Water pollutants (Class I and II) and equivalent weight of pollutant	81
Table 6.5	Basis and rate of the environmental protection tax stipulated in the EPT law	83
Table 7.1	Provisions of the relevant laws and regulations on the target responsibility system of environmental protection and performance evaluation system	87
Table 7.2	Targets of environmental protection set forth in the normative documents and the FYPs	89

List of Tables

Table 7.3	Target setting of $PM_{2.5}$ concentration in Beijing	91
Table 7.4	Evaluation guidelines and methods for the target responsibility system of environmental protection	92
Table 8.1	Solid waste treatment unit and disposal capacity in China. . . .	102
Table 10.1	The emissions targets of 12 indicators during the 9th FYP.	116
Table 10.2	The emissions targets of 6 indicators during the 10th FYP.	117
Table 10.3	Emissions targets of 2 binding indicators during the 11th FYP.	118
Table 10.4	Emissions targets of 4 binding indicators during the 12th FYP.	119
Table 15.1	List of official documents concerning cleaner production at different stages.	172
Table 16.1	Total amount and availability of natural resources in China.	184
Table 16.2	Major areas of ECM implementation in China.	185
Table 16.3	Major resource taxes listed in the resource tax ordinances. . . .	187
Table 16.4	Summary of large-scale ecological engineering projects in China.	191
Table 16.5	Ecological compensations for forest.	195
Table 16.6	Ecological compensation for coal mining in various areas. . . .	196
Table 16.7	Compensation programs for watershed conservation in Yunnan Province.	201
Table 18.1	Major responsibilities of water pollution prevention and control for various governmental agencies.	226
Table 18.2	Major responsibilities of water resources management for various governmental agencies.	227
Table 18.3	Major shifting characteristics of water governance strategies in the Water Pollution Law.	230
Table 19.1	Major responsibilities of soil pollution prevention and control for various governmental agencies.	238

List of Boxes

Box 3.1	Major supporting schemes of environmental policy	30
Box 4.1	Article 6 of the EP Law (on Trial)	38
Box 4.2	Sensitive area defined in the EIA Catalogue	45
Box 4.3	Contents of an EIA report regulated in the EIA Law	47
Box 4.4	Contents of the PEIA report for a plan and the EIA chapter for an integrated plan	56
Box 4.5	Some provisions concerning the certificate system for EIA consultancy	59
Box 6.1	Some decrees and regulations related to the emissions charges of sulfur dioxide	71
Box 6.2	Major problems within the Emissions Charges System	76
Box 7.1	Major supporting schemes of environmental policy	86
Box 8.1	Major tasks of comprehensive governance of rural environment in the Action Plan for Water Pollution	99
Box 9.1	Emission behaviors of the polluter resulting in ceasing operation	112
Box 9.2	Emission behaviors of the polluter resulting in canceling emission permit	113
Box 10.1	Various measures of statistics, monitoring and evaluation on emissions cap	118
Box 10.2	Main content of analysis and assessment on the emissions cap system during the 11th FYP	122
Box 11.1	Key tasks specified in the Air Pollution Prevention and Control for Key Regions during the 12th FYP	128
Box 12.1	Detail requirements of public participation specified in the Notice on Strengthening the Management for EIA for Construction Projects Loaned by the International Financial Organizations (the IFO)	138

Box 12.2	Provisions of public participation specified in the EIA Law (2002).	139
Box 12.3	Provisions concerning public participation in various laws and regulations.	140
Box 12.4	Major schemes of the representative of the People's Congress to supervise the environmental protection work of the government.	144
Box 12.5	Major means of cooperation and collaboration between the CCCPC and democratic parties	145
Box 13.1	Key schemes proposed by during the third plenary session of the 18th CCCPC	152
Box 13.2	Work schedule of demarcating ecological redline.	153
Box 13.3	Three redlines for water resources	154
Box 14.1	Major steps and methods of natural resource damage assessment in America.	163
Box 15.1	Detailed procedures of MCPA.	176
Box 17.1	Some documents concerning air pollution prevention and control since the 21st century.	208
Box 17.2	Several plans, schemes and programs of air pollution prevention and control since the 21st century.	209
Box 17.3	Ten tactics of the Action Plan for Air Pollution Prevention and Control	215
Box 18.1	The classifications of surface water quality.	220
Box 18.2	The classifications of ground water quality.	222
Box 18.3	Some emission standards of water pollutants	224
Box 18.4	Ten tactics of the Action Plan for Water Pollution Prevention and Control	225
Box 19.1	Ten tactics of the Action Plan for Soil Pollution Prevention and Control	243

Chapter 1
Administration on Environmental Management

In the 1950s, a series of earth shaking environmental pollution incidents occurred in the West resulting in severe damage and loss to society at large. On the subsequent, burgeoning of large-scale environmental movements has compelled governments to actively envisage environmental affairs and to be fully involved in the management of environmental issues. As a result, environmental management has gradually become an important function of government, undertaken either by existing government agencies or by specially established agencies.

China's current environmental protection management system was established in accordance with the *Constitution of the People's Republic of China* (the Constitution) and the *Law of the People's Republic of China on Environmental Protection* (EP Law). At the central government level, the State Council plays a leading role in directing and administering environmental protection programs, while the Ministry of Ecology and Environment (the MEE) carries out the actual supervision and administration of the national environmental protection programs. Other ministries and administrations also play significant roles in regulating the environmental affairs within their scope of responsibilities. Finally, local governments at all levels are responsible for environmental issues within their respective jurisdictions. Various administrative departments play different roles in environmental supervision and management.

1.1 The History of the MEE

The construction and development of environmental protection agencies in China was initiated in the 1970s. In 1971, the Leading Group of The Utilizations of Three Wastes (exhaust gas, wastewater and solid waste) was established by the National Planning Committee. In 1973, the State Council held the First National Conference on Environmental Protection, in which the setting up of an administration and management agency for environmental protection was proposed. In October 1974, the Leading Group of Environmental Protection (the LGEP) was formally founded under

© Chemical Industry Press and Springer Nature Singapore Pte Ltd. 2020

the auspices of the State Council to mark the commencement of the formation of China's environmental protection agencies. The main responsibilities of the LGEP were to institute policies, principles, and guidelines for environmental protection; to review and approve national environmental protection plans; and to organize, coordinate, supervise and inspect the implementation of environmental protection within relevant departments and at all levels. Directly under the LGEP, an executive Environmental Protection Office was established and was to be responsible for routine work. In 1978, various environmental protection institutions, namely, environmental protection bureaus (EPBs) or environmental protection offices (EPOs), were launched in municipalities directly under the Central Government, autonomous regions, and industrial cities.

In 1982, following a decision of the 23rd Meeting of the Standing Committee of the 5th National People's Congress (NPC), the Ministry of Urban and Rural Construction and Environmental Protection (MURCEP) was founded by incorporating the State Construction Commission, the State Administration of Urban Construction, the State Administration of Construction Projects, the National Administration of Surveying and Mapping, and the LGEP.

In May 1984, the Environmental Protection Commission (EPC) was established directly under the State Council to deliberate and finalize relevant policies and guidelines for environmental protection; to formulate environmental plans; and to lead, organize, and coordinate national environmental protection efforts. One of the Vice Premiers of the State Council also functioned as the Chairman of the Environmental Protection Commission, which operated through its office under the MURCEP. In December 1984, the Environmental Protection Bureau under the MURCEP was reshuffled into the National Environmental Protection Administration (NEPA), which was still under the leadership of both the MURCEP and the Environmental Protection Commission of the State Council, to be mainly responsible for the planning, coordination, supervision, and guidance of national environmental protection efforts.

In July 1988, the responsibilities for environmental protection were transferred from the MURCEP to the NEPA, which was then defined as an affiliated institution of the State Council (sub-ministerial level) and the office of the Environmental Protection Commission of the State Council to function as a competent department under the State Council for integrated environmental management. In June 1998, the NEPA was upgraded to the State Environmental Protection Administration (SEPA; ministerial level), as an affiliated institution of the State Council taking charge of the environmental protection effort. The Environmental Protection Commission of the State Council was revoked, and in July 2008, the SEPA was elevated to the Ministry of Environmental Protection (MEP). As an integral department of the State Council, the MEP is the central competent authority for environmental protection and management, with its major functions being the establishment of an integrated environmental protection system and the management and monitoring of environmental pollution prevention and control. In March 2018, the MEP was then reorganized as the Ministry of Ecology and Environment (MEE) as a consequence of the institutional reform of the State Council. The evolutionary process of the administrative

1.2 Structure and Functions of the MEE

Table 1.1 The evolutionary process of environmental protection administration in the central government

Date	Name of the institute
October 1974	The Leading Group of Environmental Protection of the State Council
May 1982	The Ministry of Urban and Rural Construction and Environmental Protection
May 1984	The Environmental Protection Commission of the State Council
December 1984	The National Environmental Protection Administration (under the MURCEP and the Environmental Protection Commission)
July 1988	The National Environmental Protection Administration, the NEPA (sub-ministerial level)
June 1998	The State Environmental Protection Administration, the SEPA (ministerial level)
July 2008	The Ministry of Environmental Protection, the MEP
March 2018	The Ministry of Ecology and Environment, the MEE

department for environmental protection under the State Council is conceptually illustrated in Table 1.1.

1.2 Structure and Functions of the MEE

The MEE is located in Beijing and is made up of a number of functional departments and institutes, together with 12 dispatched agencies located in various regions, as shown in Fig. 1.1. In addition, there are 23 affiliated public institutions (including scientific research institutes, publishing agencies, monitoring departments, and so on) and five social groups (including academic associations, environmental associations, foundations, etc.) under the direct command of the MEE, as listed in Table 1.2. The key functions of the MEE can be summarized as follows: formulation of environmental protection regulations, enactment and supervision of environmental planning for key regions and river basins, adjudication of major environmental pollution accidents and ecological damage incidents, and management of other administrative activities.

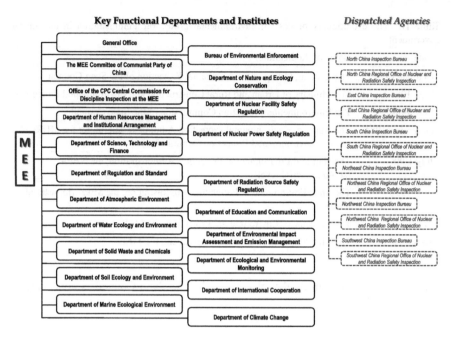

Fig. 1.1 Structure of the ministry of ecology and environment

Table 1.2 Affiliated public institutions and social groups directly led by the MEE

#	Name
Public institutions	
1	Center of Environmental Emergency and Accident Investigation
2	Agency Service Center
3	Chinese Research Academy of Environmental Science
4	China National Environmental Monitoring Centre
5	Sino-Japan Friendship Centre
6	Policy Research Centre of Environmental and Economy
7	China Environmental News
8	China Environment Publishing Group
9	Nuclear and Radiation Safety Centre
10	Foreign Economic Cooperation Office
11	Nanjing Institute of Environmental Science
12	South China Institute of Environmental Science

(continued)

1.2 Structure and Functions of the MEE

Table 1.2 (continued)

#	Name
13	Chinese Academy for Environmental Planning
14	Appraisal Center for Environment & Engineering
15	Satellite Environment Centre
16	China-ASEAN Environmental Cooperation Centre
17	China Centre for SCO Environmental Cooperation
18	Lancang-Mekong Environmental Cooperation Centre
19	Solid Waste and Chemicals Management Center
20	National Centre for Climate Change and International Cooperation
21	Beijing Conference and Training Base
22	Xingcheng Environmental Management Research Center
23	Beidaihe Environmental Technology Exchange Center
Social groups	
1	Chinese Society for Environmental Sciences
2	China Environmental Protection Foundation
3	China Environmental Culture Promotion Association
4	China Forum of Environmental Journalists
5	Chinese Ecological Civilization Research and Promotion Association

Chapter 2
Evolution of China's Environmental Policy

Environmental protection movements in China were not initiated until the 1970s. Though ecological degradation and environmental pollution had gradually emerged, locally and regionally, along with rapid economic growth and drastic social development, not much attention was paid and concrete environmental protection measures were rare. The development of China's environmental protection measures received a great boost, with its participation in the United Nations Conference on the Human Environment held in Stockholm in June 1972, following which the Chinese government began to institute policies and guidelines for national environmental protection.

2.1 In the 1970s: The Principles of National Environmental Protection

During the United Nations Conference on the Human Environment, eight principles for environmental protection, namely, overall planning, rational layout, comprehensive utilization, turning disadvantage into advantage, relying on the public, all-people participation, protecting environment, and benefiting the people, were first proposed by China. In August 1973, these eight principles, known as the '32-Chinese-Characters Principles', were approved as the guidelines for environmental protection by the First National Conference on Environmental Protection held in Beijing, and were reaffirmed in the first national document of China's environmental protection, *Several Regulations Concerning Protecting and Improving Environment (on Trial Draft)*, and the *Law of the People's Republic of China on Environmental Protection (on Trial)*, promulgated on September 13, 1979. Moreover, in these eight principles, the objectives and basic measures for environmental protection were clearly clarified, the general principles and direction for the implementation of environmental protection stipulated, and some key aspects and major issues of environmental protection identified. During the 1970s, a series of environmental management schemes were instituted and implemented according to these eight principles, along with some regulations, provisions, and management measures that served as the concretization and extension of these eight principles. Despite some deficiencies and restrictions, practical experience has demonstrated that these eight principles have played an essential role in promoting China's environmental protection for quite a long time, as they were fully in compliance with China's national conditions at that time.

2.2 In the 1980s: The Principles of Three-Simultaneity and Three-Unification

With the initiation of the Reform and Opening-Up Policy in December 1978, economic reform has been steadily strengthened, and the political and economic situations have both undergone great changes since the 1980s. In addition, China's environmental protection has experienced significant changes due to gradually emerging and exacerbating environmental problems and an increasing awareness and understanding of environmental issues. Environmental protection regulations and policies were primarily established through enhancing administrative schemes of environmental governance. Environmental protection has become one of the basic national policies of China by strongly promoting the advancement of legal institutions, capacity building and administrative governance for environmental protection.

Based on the practical experience of environmental protection work summarized since the First National Conference on Environmental Protection, an environmental protection strategy, 'economic construction, urban-rural construction, and environmental construction should be planned, implemented and developed synchronized so as to realize the unification of economic benefits, social benefits and environmental benefits', was then proposed during the Second National Conference on Environmental Protection in December 1983 and came to be known as the "Three- Simultaneity and Three-Unification". This strategy was a significant advancement over the 32-Chinese Characters Principles. Based on the summary of China's historical progress, in conjunction with its national conditions and practical experience of environmental protection, it indicated the right path toward resolving the environmental problems and marked a major improvement in the philosophies of environmental management and in the guidance offered for the development of new theories of environmental management in China.

The most prominent feature of China's environmental management system, preliminarily established in the 1980s, was the strengthening of administrative management. This can be illustrated by the following quote from 1989 by Geping Qu, who is honored as the Father of China's Environmental Protection: "At that time, it was generally recognized by the decision-making level that there was not enough money to be invested in environmental governance. In addition, many environmental problems mainly resulted from poor management. Many environmental problems can be resolved through the enhancement and improvement of management." According to the accepted wisdom and conditions prevailing at that time, the most practical and effective method of implementing environmental protection was to employ administrative and regulative schemes of mandatory management and control by government. Thus, to strengthen environmental management, measures such as supervising environmental governance through administrative directives and promoting environmental protection through monitoring, has become the dominant policy strategy, which proved to be very effective through implementation at that time.

2.3 In the 1990s: Sustainable Development

In the 1990s, environmental laws and regulations were further revised and enhanced to significantly promote environmental governance in key regions while the advantages of administrative management gradually emerged. In 1991, the objectives of environmental protection plan were integrated into the Eighth Five-Year Plan of National Economic and Social Development (the 8th FYP), for the first time. In August 1992, implementation of a sustainable development strategy was explicitly stipulated by the Chinese Central Government in the proposed *Ten Countermeasures of Environment and Development*, which was the first guidance document for environment and development. Furthermore, in the same year, for the first time, the environmental statistics data were incorporated into the Annual Statistical Bulletin of National Economic and Social Development, which embodies the breakthrough and achievements of the environmental protection movement for nearly twenty years. The sustainable development strategy denotes the substantial transformation from the principles of Three-Simultaneity and Three-Unification that dominated the 1980s. Environmental protection and social and economic development were regarded as an integrated system in the sustainable development strategy meant to typify the environmental concerns and to reflect the latest recognition of the essence of social development by the Chinese Central Government into the directorial objectives and strategic guidelines of national development.

To improve the legal system of environmental protection, several laws and regulations were amended [including the *Law of the People's Republic of China on Water Pollution Prevention and Control*, (the Water Pollution Law) the *Law of the People's Republic of China on Air Pollution Prevention and Control* (the Air Pollution Law), and the *Law of the People's Republic of China on Marine Environment Protection* (the Marine Protection Law)] or newly promulgated [including the *Law of the People's Republic of China on Solid Waste Pollution Prevention and Control* (the Solid Waste Pollution Law), the *Law of the People's Republic of China on Environmental Noise Pollution Prevention and Control* (the Noise Pollution Law), and the *Law of the People's Republic of China on Land Desertification Prevention and Control* (the Land Desertification Law)] by the Environmental Resources Committee of the NPC, successively. For capacity building in environmental protection institutes, the NEPA was upgraded as an affiliated institution of the State Council (vice-ministerial level) in 1993 and promoted to the SEPA (ministerial level) in 1998, to greatly advance the status of environmental protection at the national level, and to enhance policy support and resource allocation in environmental protection. In the area of environmental management, through continuous enhancement of administrative management, three key areas of environmental governance were identified, including water pollution governance of three rivers (Huaihe, Haihe, and Liaohe) and three lakes (Taihu, Chaohu, and Dianchi), air pollution governance in two-control-zones (acid rain control zone and sulfur dioxide control zone), and integrated governance in the spheres of industrial pollution and urban environment.

2.4 In the 2000s: Scientific Outlook on Development

At this stage, the Chinese government commenced a comprehensive form of environmental governance through the integration of all kinds of power and resources and the enactment of a number of significant environmental protection initiatives. In addition, various policy tools, such as environmental taxation, green finance and so on, were designed and promoted.

From the perspective of legal institutions, it was explicitly specified in the Fifth National Conference on Environmental Protection in January 2002, that environmental protection is an important government function, and the society at large should be devoted to environmental protection according to the requirements of the socialist market economy. In October 2003, the concept of a scientific outlook on development was set forth in the Third Plenary Session of the Sixteenth Central Committee of the Communist Party of China (CCCPC). In October 2005, the guidelines to build a resource-conserving and environment-friendly society were put forward in the Fifth Plenary Session of the Sixteenth CCCPC. In April 2006, it was proposed in the Sixth National Conference on Environmental Protection that environmental protection should be given a more strategic position. In October 2007, the formation of an ecological civilization society was advocated in the Seventeenth National Congress of the Communist Party of China (NCCPC), along with the notion of industrial structure, growth modes, and consumption patterns for resource saving and ecological protection. Furthermore, several laws and regulations were promulgated, including the *Law of the People's Republic of China on Promoting Cleaner Production* (Cleaner Production Promotion Law), the *Law of the People's Republic of China on Environmental Impact Assessment* (EIA Law), the *Law of the People's Republic of China on Radioactive Pollution Prevention and Control* (Radioactive Pollution Law), the *Law of the People's Republic of China on Renewable Energy* (the Renewable Energy Law) and the *Law of the People's Republic of China on Promoting Circular Economy* (Circular Economy Promotion Law). As stated earlier, in July 2008, the MEP (promoted from the SEPA), an integrated department of the State Council for environmental protection and management, became an extremely important functional institute responsible for the overall planning and coordination in decision-making, planning and management of major environmental issues.

2.4 In the 2000s: Scientific Outlook on Development

In the context of rapid deteriorating environmental conditions, in order to increase the effectiveness of environmental policies, diverse financial and taxation schemes and preferential investment policies (i.e., credit and loans) were implemented by various government departments and agencies. In 2002, the Franchise System was initiated to allow private and foreign capital to be invested in the fields of sewage treatment, waste management, heating and other fields. Subsequently, several incentives and price policies in favor of environmental protection were employed, such as electricity price compensation policies for thermal power plants with desulfurization facilities in 2004, with denitrification facilities in 2011, and so on.

In respect of taxation, enterprises known for being "energy-saving and environment-friendly" became eligible for income tax exemption for the first three-years and 50% income-tax deduction for the subsequent three-year period. Enterprises involved in wastewater treatment, reclaimed water and waste management industries are entitled to value-added tax exemption or instant refund. Products from desulfurized processes are entitled to half value-added tax deduction. Investment on environmental protection equipment and instruments can be deducted from income tax. With regard to investment, the Green Credits policy, initiated since July 2007, required that state-owned banks and commercial banks should provide substantial support to green industries. The implementation of the above-mentioned financial and taxation schemes, credit and loans, and other preferential investment policies has initiated a new market-oriented path for China's environmental governance to not only promote the construction of environmental infrastructure but also actuate more social resources to be invested in environmental protection, which has achieved remarkable environmental benefits.

In short, the environmental protection legislation in this stage was unprecedented and administrative regulations were launched intensively to establish a strong structure of control and management. In particular, the enactment and implementation of the EIA Law reflects the significant change in the government's attitude toward environmental protection, which represents the transition from the recognition of "pollution first, treatment later" to the rigid requirement of "evaluation first, construction later." This is a critical but beneficial move toward paying more attention to the principle of prevention first. Subsequently, a number of major events occurred in 2004, 2006 and 2007, whereby investment totaling hundreds of billions in various industries was suspended or cancelled due to their inability to meet the requirements of the environmental impacts assessment (known as "Huan-Ping-Feng-Bao" in Chinese), which demonstrates the determination of the government to strictly implement environmental governance policy. In addition, the application of financial and taxation environmental policy has demonstrated a new approach of attracting more capital to be invested into environmental governance.

2.5 In the 2010s: Ecological Civilization

After 40 years of practice in environmental protection, with the continuous evolution of policy orientation, economic structure and social conditions, a comprehensive transformation of environmental governance in China is inevitable. Starting with its enrollment as one of China's national strategies, to the perfection of environmental protection legislation and the implementation of an environmental protection system, to environmental protection supervision and management and public participation, and subsequently leading to the rapid growth of social forces in environmental protection, the development of environmental protection in China has reached a new era, with a more scientific and specialized institutional system, more innovative policy tools, and more wisdom and support from society at large.

During the 18th NCCPC held in November 2012, it was proposed that ecological civilization construction should be placed at a prominent strategic level to be integrated with the other four constructions, namely, economic construction, political construction, cultural construction, and social construction. Ecological civilization is not only an important conceptual innovation but also a key national governance strategy for China. As portrayed in the *Opinions on Accelerating the Advancement of the Ecological Civilization Construction* and the *General Planning for the Institutional Reform of Ecological Civilization*, China's vision of ecological civilization is "Enjoying a beautiful home with a blue sky, green land, and clean water is articulated as the dream of every Chinese and thus placed at the core of the Chinese Dream", which is embedded in the idea of building a "Beautiful China". During the Third Plenary Session of the 18th CCCPC held in November 2013, it was proposed that a systematic and comprehensive ecological civilization system be established, along with the introduction of certain notions, including property rights to resources, ecological redline, and so on. As stipulated in the *Opinions on Accelerating the Advancement of the Ecological Civilization Construction*, promulgated by the CCCPC and the State Council in May 2015, specific objectives of ecological civilization cover spatial development, resource use, ecological and environmental quality, and regulatory systems, as shown in Table 2.1.

In April 2014, the EP Law Amendment was approved during the 8th Meeting of the Standing Committee of the 12th NPC, in which the government's environmental responsibilities were explicitly demarcated; enterprises' and operators' environmental liabilities were meticulously elucidated; and citizens' environmental rights, obligations and duties were specifically stipulated. Moreover, in the EP Law (2014), the efficacy of the environmental economic policies was further confirmed, the procedures of environmental public interest litigation clearly specified, and the degrees of penalties and liability for violating emission standards greatly elevated.

In July 2015, in the *Measures of Public Participation in Environmental Protection* promulgated by the MEP, concrete measures to implement public participation in environmental protection were specified. In August 2015, in the *Measures of Pursuing the Liability of Party and Government Leading Cadres for Ecological and*

2.5 In the 2010s: Ecological Civilization

Table 2.1 The targets of ecological civilization by 2020

Targets	Contents
Further optimize spatial development	• Implement "Principle Functions Zoning" • Delineate and enforce ecological redlines • Enforce redlines for protecting cultivated land
Utilize resources more efficiently	• Reduce CO_2/GDP by 40–45% over 2005 level • Reduce energy intensity further • Increase resource productivity greatly • Confine total water consumption under 670 billion cubic meters • Limit water consumption/CNY 10×10^3 of industrial added value under 65 cubic meters • Increase irrigation efficiency ("effective utilization coefficient of farmland irrigation water") above 0.55 • Increase non-fossil energy to 15% of primary energy consumption
Improve the overall quality of ecological environment	• Reduce total discharge of sulfur dioxide (SO_2), nitrous oxide (NO_X), chemical oxygen demand (COD) and ammonia nitrogen (NH_3–N) • Improve air quality and water quality of key watersheds and offshore areas • More than 80% of the key rivers/lakes/water functional areas meeting water quality standards • Improve safety and security of drinking water • Stabilize overall soil quality • Control environmental risks • Increase forest coverage over 23% • Increase prairie's vegetation coverage over 56% • Maintain minimum wetland areas at 533,333 km^2 • Reclaim more than 50% of the reclaimable desert • Preserve the natural shorelines • Control biodiversity loss and enhance stability of nation-wide ecosystems
Establish major regulatory systems of eco-civilization	• Shaping up an Eco-civilization system characterized by "source prevention, pollution process control, damages compensation, and accountable liability" • Remarkable achievements in constructing critical systems such as the ownership and use of natural assets, ecological redlines, compensation for ecological protection, and ecological and environmental protection management

Source State Council (2015)

Environmental Damage (on Trial) jointly issued by the CCCPC and the State Council, the responsibilities of environmental protection and resources conservation for local party committees and governments at all levels were detailed so as to stringently pursue the liability of any dereliction of duty. In October 2015, in the *Recommendations on Enacting the Thirteenth Five-Year Plan of National Economic and Social Development* ratified by the 5th Plenary Session of the 18th CCCPC, it was proposed that a vertical management system should be implemented for environmental monitoring, supervision, and enforcement at the provincial level. In December 2016, the *Index System of Green Development* and the *Objectives System of Ecological Civilization Construction Auditing* were jointly enacted by the National Development and Reform Commission (the NDRC), the National Bureau of Statistics (the NBS), the MEP, and the Organization Department of the CCCPC.

The promulgation and implementation of the above-mentioned decrees and provisions have provided diverse compulsive mechanisms and effective tools for the practice of environmental protection in China, for example, more systematic institutional systems, more stringent supervision and execution, more diversified participation, and so on. Through multilevel governance transformation (including institutional construction, implementation according to laws, public participation, and multivariate shared-governance), environmental governance in China has currently become a symbol of the modernization of the national governance system, with significant enhancements to its innovation, efficiency, evolution, and advancement.

2.6 Concluding Remarks

All environmental issues, our understanding of them, and the relevant policies and corresponding countermeasures are the result of prevailing historical conditions, economic structure, and social development. The changing trend of environmental policy is strongly related to the specific political process, economic structure, and social development to reveal the levels and limits of understanding of environmental issues at a particular point in time, with careful consideration to historical background and limitations of contemporary circumstances.

Since the 1970s, China's environmental policy has gradually developed the capacity to adapt to contemporary social and economic development, through constant evolution, development, improvement, and perfection. Only under the larger framework of historical evolution, can the recognition and analysis of environmental policy be performed to acquire an objective and distinct understanding. Furthermore, the essence of the larger framework of historical evolution is actually inherent in the rapid transformation of economic structure and severe societal change in China for the past 50 years. Apparently, China's environmental policy has always been an important integral part of the Chinese political, social and economic system. Currently, China's socio-economics is undergoing a transition from a planned economy to a market economy, where environmental policy is only one of many important aspects. The development of China's environmental policy is basically a top-down

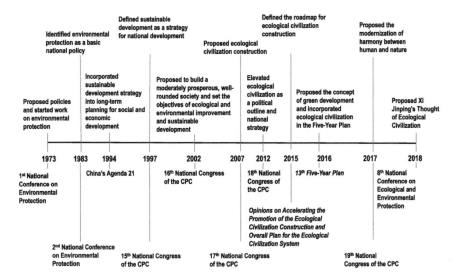

Fig. 2.1 The development process of China's environmental policy (1973–2018)

process, initiated, designed, improved, and enhanced by the central government, with very little influence from external forces and other social factors, as illustrated in Fig. 2.1.

Reference

State Council. (2015). Opinions on Accelerating the Construction of Ecological Civilization. www.gov.cn/gongbao/content/2015/content_2864050.htm (in Chinese). [2019-03-22].

Chapter 3
Environmental Planning

Environmental planning is the process of making appropriate time and spatial arrangements relating to people's activities and environments according to social and economic laws, ecological and geographical theory in order to achieve sustainable development (Jin, 2014). Environmental planning endeavors to manage these processes in an effective, orderly, transparent, and equitable manner for the benefit of all constituents within such systems for both the present and the future. It is a broad field, bridging the disciplines of geology, environmental science, ecology, law, economics, and public policy.

3.1 The Development of Environmental Planning

In the early stage of environmental protection work, "comprehensive planning" and "rational layout" were included in the guidelines for environmental protection work, which were put forward during the First National Conference on Environmental Protection held in 1973. In October 1974, the goal, "to control pollution within 5 years and to resolve pollution issues with 10 years," was proposed by the State Council's LGEP to guide the institution of pollution governance planning, nationwide. However, in reality, this goal proved to be unrealistic due to the lack of a comprehensive and profound understanding of the objective conditions of environmental protection and a clear recognition of the relationship between environmental planning and socio-economics development planning at that time.

Since the Third Plenary Session of the 11th CCCPC held in 1978, it has been gradually accepted by the policy makers that environmental protection must be incorporated into economic development, and that economic development should take place in tandem with environmental protection. Therefore, while compiling the 6th FYP, environmental protection was then incorporated as an essential part of the 6th FYP, for the first time, to institute "strengthening environmental protection, preventing further environmental quality deterioration from environmental pollution and improving environmental quality in some key areas" as the basic functions of environmental protection.

At the end of 1983, environmental protection was determined as one of the basic national policies during the Second National Conference on Environmental Protection, convened in Beijing to propose "the synchronization of planning, implementation and development for economic, urban-rural and environmental constructions, and the unification of economic, social and environmental benefits" as the strategic guidelines, and to promulgate "prevention first", "polluter pays", and "strengthening environmental management" as the three major policies of environmental protection. The principles of pollution prevention-control and the responsibilities of economic departments and agencies regarding environmental governance were clear stipulated in these guidelines and policies to promote the development of regional environmental planning and enterprise environmental planning.

In October 1985, the Working Meeting on National Urban Environmental Protection was held in Luoyang (Henan Province) by the Environmental Protection Committee of the State Council to put forward the guidelines for integrated urban environmental protection, to illustrate the route of comprehensive governance on urban environmental protection, and to define the Mayor's responsibilities regarding urban environmental quality. In addition, as clearly pointed out in this Meeting, a plan of comprehensive governance for urban environmental protection was expected to be instituted that aimed to include major tasks, such as the promotion of a virtuous circle of ecological environment through pollution prevention and control, and the improvement of the ecological environment; the utilization of urban resources as a best practice; and the provision of clean, healthy, comfortable and elegant living and working environments with minimum effort and consumption to all citizens.

In December 1989, the EP Law (1989) was promulgated. As stipulated in Article 10, an environmental protection plan had to be incorporated into the Plan for the National Economic and Social Development. Moreover, according to Article 12, the competent authorities for environmental protection at the county level and above were expected, in conjunction with the relevant departments, to investigate and evaluate the environmental conditions within their jurisdictions so as to formulate environmental protection plan that would be revised carefully by the Planning Department and then submitted for approval to the People's Government at the same level for implementation. Hence, the agencies and procedures for the institution of an environmental protection plan were clearly defined in the EP Law (1989).

In August 1992, the *Ten Strategies on Environment and Development* was approved and disseminated by the CCCPC and the State Council. As pointed out in Article 1: Implementing a Sustainable Development Strategy: "In order to speed up economic development and to solve environmental problems, it is essential to transform the development strategy and to adopt the path of sustainable development. Therefore, it is necessary to reiterate the strategic guideline, 'the synchronization of planning, implementation and development for economic, urban-rural and environmental constructions'. The People's Government at all levels and the relevant departments must compile an environmental protection plan, while instituting and implementing the development strategy."

In 1993, the *Compiling Technical Outlines of the Plan for Comprehensive Urban Environment Governance* was released by the NEPA, requiring all cities to formulate a Comprehensive Urban Environment Improvement Plan; subsequently a tool book, the *Handbook for Environmental Planning*, was published in November 1994.

In March 1996, the *Outlines of the Ninth FYP for the National Economic and Social Development of the People's Republic of China and the Long-Term Objectives for 2010* was approved by the Fourth Session of the Eighth NPC to specifically put forward the implementation of a sustainable development strategy and the goal of trans-century environmental protection. Since the 9th FYP, ecological protection and rural environmental protection were integrated into the environmental protection plan and the Total Emission Cap Plan, Key Projects Plan and Key Watersheds Plan and Regional Plan were formulated. In addition, with the imminence of the 21st century, environmental problems of and environmental protection plans for the 21st century were then becoming the new hotspots of research in environmental planning.

In early 2000, the *Compiling Technical Outlines of the Local Environmental Protection Plan during the Tenth FYP and the Long-Term Objectives for 2015* was instituted by the SEPA. On March 15, 2001, the *Outlines of the Tenth FYP for the National Economic and Social Development of the People's Republic of China* was approved by the Fourth Session of the Ninth NPC, where in a people-oriented ideology and the principle of equal importance between pollution prevention and ecological protection were solicitously embodied. Some guiding ideologies were also proposed, such as employing the total emission cap as the principal axis; recognizing regional pollution prevention and control measures containing anthropogenic ecological destruction as the key points; and embracing government leadership, promotion by enterprises, and public participation as the direction for mechanism innovation. Further, the *Plan for the Disintegration of Total Emission Cap on Major Pollutants during the Tenth FYP* was publicized as an appendix of the *National Environmental Protection Plan during the 10th FYP* in December 2001, where total emission cap for six major pollutants were defined and allocated into various targets for provinces, autonomous regions, municipalities, and designated cities. In addition, based on 14 years' experience in water environment function zoning across the country, the national integration of water environment function zoning was preliminarily compiled by the SEPA. It was the first instance of systematic and nationwide water environment function zoning. Furthermore, ecological function zoning for 31 provinces, autonomous regions, and municipalities was also primarily completed. Thus, it represented a great advancement in environmental protection plans for key areas. Moreover, the *Environmental Protection Plan for the Pearl River Delta Region* (the first regional environmental protection plan through legislation) and *the Environmental Protection Plan for Guangdong Province* were approved by the People's Congress of Guangdong Province in 2002 and 2005. Later on, the compilation of an environmental protection plan for the Yangtze River delta region and the Beijing-Tianjin-Hebei region were initiatively deployed in succession.

During the period of the 11th FYP, environmental protection was recognized as the key connotation to carry out scientific development, the major means to transform economic development patterns, and the fundamental measure to promote the construction of ecological civilization, by imposing total emission caps on two major pollutants, COD and SO_2, as the binding indicators for national economic and social development. For the first time, the planning concepts of the Environmental Protection Plan during the 11th FYP were reported to the Standing Committee of the Politburo of the CCCPC and the Standing Committee of the State Council. In addition, the national environmental protection plan was promulgated by the State Council for the first time to denote that the substances and requirements for environmental protection and related areas were increasingly becoming one of the for government's core tasks at all levels.

During the period of the 12th FYP, the strategic thinking of "developing while protecting, protecting while developing," was concretely embodied in the formulation of an environmental protection plan to promote environmental protection planning at a deeper development stage. In 2014, the EP Law (2014) was amended and proclaimed to clearly specify that environmental protection tasks would be incorporated into regional economic and social development plans for all people's governments at the county level or above. According to the national economic and social development plan, the national environmental protection plan would be formulated by the competent department for environmental protection under the State Council, in conjunction with relevant departments, and reported to the State Council for approval. For local people's governments at the county level or above, the competent departments for environmental protection would be expected to formulate local environmental protection plans according to the requirements of the national environmental protection plan, in conjunction with the relevant departments, and report to the people's governments at the same level for approval. The goals, tasks and safeguard measures of ecological protection and pollution prevention and control had to be included in the content of the environmental protection plan, which would be linked with the key function zoning plan, general land use plan, urban-rural plan, etc.

Since the 18th NCCPC held on November 8, 2012, ecological civilization was promoted as the national strategy and ecological environmental protection became a priority item on the agenda for the CPC and the Central Government. As repetitively emphasized by General Secretary Jinping XI, "Green hills and clear waters are as good as mines of gold and silver. Protecting the ecological environment is the same as protecting your eyes and treating ecological environment is the same as treating your life." As pointed out by Premier Keqiang LI, "We need to enhance environmental governance and to plough the win-win route by improving environmental quality while promoting economic development, resolutely." The notion of ecological civilization, respect for nature, conforming to nature, and conservation of nature became the guiding principles for the 13th FYP. As stipulated in the *Ecological and Environmental Protection Plan during the 13th FYP*, the main targets to be achieved by 2020 are an overall improvement in environmental quality, development of a green lifestyle and production techniques, advancements in developing a low-carbon economy, huge reductions in emissions of major pollutants, effective and efficient control over environmental risks, control over biodiversity loss, enhancement of ecosystem stability, establishment of ecological safety barriers, significant progress in the modernization of the national governance system and capacity for

Table 3.1 Categories of environmental protection plan

Classification	Details
By administrative jurisdictions	National environmental protection plan
	Provincial environmental protection plan
	Municipal environmental protection plan
	Environmental protection plan at county level
By regions	Regional environmental protection plan (cross-administrative regions)
	Watershed environmental protection plan
By sectors	Environmental protection plan for industrial sector, agricultural sector, transportation sector, and etc.
By environmental elements	Pollution prevention and control action plan for air pollution, water pollution, soil pollution, and etc.
	Ecological protection and construction plan for biodiversity, natural reserve, forest, water source, and etc.
Other plans related to ecological environmental protection	Circular economy planning, ecological city planning

ecological environment, and the conformation of the stage of ecological civilization construction to the goals of building a comprehensive moderately prosperous society.

At present, the environmental protection planning system has been fundamentally established to cover most areas of environmental protection, as shown in Table 3.1. Though there is no specialized law of environmental planning, the requirements of environmental planning are explicitly described in several laws and regulations concerning resources conservation and environmental protection, as shown in Table 3.2.

3.2 Environmental Protection and the Plan of National Economic and Social Development

The initial process of incorporating the environmental protection plan into the plan of national economic and social development has been rather lengthy. Due to historical reasons, there was no environmental protection plan during the first five FYPs.

Table 3.2 Contents related to environmental planning in selected environmental law and regulations

Law and regulations	Contents
The law of the People's Republic of China on environmental protection (2014 revision)	
	Article 13 The people's governments at and above the county level shall include the environmental protection work in national economic and social development plans
	The environmental protection administrative department of the State Council shall, in conjunction with the relevant departments, develop national environmental protection plan in accordance with the national economic and social development plan, and submit it to the State Council for approval, publication, and implementation
	The environmental protection administrative departments of the local people's governments at and above the county level shall, in conjunction with the relevant departments, develop the environmental protection plans for their respective administrative regions in accordance with the requirements of the national environmental protection plan, and submit them to the people's governments at the same level for approval, publication, and implementation
	Environmental protection plan shall include the objectives and tasks of and safeguards for ecological protection and pollution prevention and control, and be connected with the major function zoning plan, land use comprehensive plan, and urban and rural development plan, among others
The law of the People's Republic of China on air pollution prevention and control (2015 revision)	
	Article 3 The people's governments at and above the county level shall include atmospheric pollution prevention and control in the national economic and social development planning and increase financial support for it
	The local people's governments at all levels shall be responsible for the quality of the atmospheric environment of their respective administrative regions, make plans, take measures, control or gradually reduce the discharge of atmospheric pollutants, and make the atmospheric environment quality reach the prescribed standards and gradually improve it
	Article 14 The people's governments of cities failing to reach the national atmospheric environment quality standards shall timely make plans for reaching atmospheric environment quality standards within the prescribed time and take measures to reach atmospheric environment quality standards within the time limit specified by the State Council or the provincial people's governments
The law of the People's Republic of China on water pollution prevention and control (2017 revision)	
	Article 4 The people's governments at or above the county level shall integrate water environment protection into the national economic and social development plan
	Article 16 The prevention and control of water pollution must be under unified planning by drainage area or region. The water pollution prevention and control plan of an important river or lake determined by the state must be prepared by the administrative department of environmental protection and departments of macroeconomic control and water administration under the State Council together with the people's government of the related province, autonomous region or municipality directly under the Central Government, and be submitted to the State Council for approval
	The water pollution prevention and control plan of a river or lake across more than one province, autonomous region or municipality directly under the Central Government, other than one prescribed in the preceding paragraph, shall be prepared by the administrative departments of environmental protection under the people's governments of the related provinces, autonomous regions or municipalities directly under the Central Government together with the competent departments of water administration at the same level …
	The water pollution prevention and control plan of a river or lake across more than one county in a province, autonomous region or municipality directly under the Central Government shall be prepared by the administrative department of environmental protection under the people's government of the province, autonomous region or municipality directly under the Central Government together with the competent department of water administration at the same level …
	Article 17 The relevant city or county people's government shall, according to the requirements of the objective of improving water environment quality as determined in accordance with water pollution prevention and control plan, make compliance plan, and take measures to reach the objective on schedule

(continued)

3.2 Environmental Protection and the Plan of National Economic and Social Development

Table 3.2 (continued)

Law and regulations	Contents
The law of the People's Republic of China on soil pollution prevention and control (2019)	
Article 11 The people's governments at and above the county level shall include the prevention and control of soil pollution in national economic and social development plans and environmental protection plans The ecological and environmental departments of local people's governments at and above the level of city divided into districts shall, in conjunction with development and reform, agriculture and rural affairs, natural resources, housing and urban-rural development, forestry and grassland, and other departments, based on the requirements of environmental protection plans, land use, soil pollution surveys, and monitoring results, among others, prepare soil pollution prevention and control plans …	
The law of the People's Republic of China on solid waste pollution prevention and control (2016 revision)	
Article 4 The people's governments at or above the county level shall incorporate the prevention and control of environmental pollution by solid wastes into their environmental protection plan and adopt economic and technical policies and measures to facilitate the prevention and control of environmental pollution by solid wastes Article 29 The relevant departments of the people's governments at or above the county level shall formulate a prevention and control plan of environmental pollution by industrial solid wastes, popularize the advanced production techniques and equipment which can reduce the discharge and harm of industrial solid wastes and promote the prevention and control of environmental pollution by industrial solid wastes Article 54 The environmental protection administrative department of the State Council shall, jointly with the economic comprehensive macro-control department of the State Council, formulate the plan for constructing facilities and sites for centralized treatment of hazardous wastes …	

3.2.1 Environmental Protection During the 6th FYP (1981–1985)

At the beginning of 1982, *Some Preliminary Suggestions concerning Strengthening the Guidance on Environmental Protection Plan* was promulgated by the LGEP Office of the State Council to assert that the environmental protection plan should ensure four balances:

- the balance between industrial emission and environmental capacity;
- the balance between construction projects and facilities for pollution prevention and control;
- the balance between urban population growth and economic and social development (including urban environment improvement);
- the balance between the exploitation and utilization of natural resources and resources conservation and the increasing utilization of renewable resources.

At the same time, several indices and measures were introduced to solicit the opinions of the provincial and municipal environmental protection departments as the preliminary preparation work for the incorporation of environmental protection into the 6th FYP. After much endeavor and hard work, environmental protection was finally incorporated into the 6th FYP as an individual chapter—a groundbreaking achievement for environmental protection.

During the 6th FYP, several tasks were set as the key control targets, including preventing the spread of new pollution, preventing the ecological environmental status from further deterioration, resolving prominent pollution problems, and improving the urban environmental status of popular tourist destination cities (including Beijing, Suzhou, Hangzhou and Guilin). In addition, six target indicators, such as the emission loads and processing capacity for industrial wastewater, the emission loads and processing capacity for hazardous gases, and the generation and comprehensive utilization of industrial solid waste, were stipulated. However, these targets were not fully realized, unfortunately, due to the fact that these environmental indices were not concretely incorporated into the annual plans of various industrial departments and local governments, and the investments in environmental protection were not included in the state budget. Nevertheless, it was a major breakthrough for the environmental protection plan to be incorporated into the national economic and social development plan and a good starting point to gradually accumulate valuable experience.

3.2.2 Environmental Protection During the 7th FYP (1986–1990)

Environmental protection was successfully incorporated into the 7th FYP, where the goals and tasks of environmental protection were more specific than those of the 6th FYP. In order to enhance the comparability and feasibility of the 7th FYP, the indices of some targets were specifically quantified. The main features of the 7th FYP could be summarized as follows:

① In addition to the five tasks, namely, "industrial pollution prevention and control", "protection of water quality of rivers, lakes, reservoirs and oceans", "major urban environment protection", "rural environment protection" and "ecological environment improvement", stipulated in the chapter of Environmental Protection (Chap. 52), environmental protection and territory governance were also included as the overall strategic objectives for the social development programs.
② Within the economic development plans of industry, agriculture, transportation, energy, water conservancy, urban construction, etc., the relevant content of environmental protection was put forward in response to the requirements. The concept of synchronization between economic, social, and environmental development was recognized and realized in the national economic and social development plan, which denotes an important measure of ideological progress for national economic planning.
③ In many provinces and cities, environmental protection was incorporated into local 7th FYPs, for the first time, to put forward the goals, tasks, and measures of environmental protection during the 7th FYP.

The incorporation of environmental protection into the 7th FYP obviously enhanced the macroscopic regulation on national environmental protection, increased the investment in environmental pollution control, and greatly promoted the work in environmental protection. Nevertheless, as environmental protection had not yet been put into the annual plan of national economic and social development, environmental protection investment could hardly be implemented. Thus, the problem of incorporating environmental protection be as one of the major courses of national economic development had not yet been completely solved.

3.2.3 Environmental Protection During the 8th FYP (1991–1995)

Since 1989, research on the compiling work of the annual environmental plan was carried out by the SEPA, along with several pilot projects in various provinces and cities, under the support of the State Planning Commission (the SPC). At the same time, during the compilation of the 8th FYP, local governments at all levels, and various competent departments for industries, agriculture, forestry, and water conservancy, were required by the SEPA and the SPC to incorporate environmental protection into their 8th FYPs, whereby concrete goals, tasks, and measures were specified and disintegrated into individual annual plan, on the basis of simultaneous development policy.

In the development strategy of the 8th FYP, the status of environmental protection as a national policy was highlighted, the strategic goals of environmental protection were proposed, a discrete chapter of environmental protection was stipulated, and the specified environmental protection plan was compiled in the 8th FYP. Within this specific environmental protection plan, quotas of 21 indicators reflecting pollution prevention and control for Total Emission Cap were allocated into 30 provinces, autonomous regions and municipalities directly under the Central Government and 14 municipalities with independent planning status. Seven environmental quality indicators were issued in 50 key cities for environmental protection. As an important part of the 8th FYP, this specific plan was issued directly to regional governments, various competent departments, and diverse industrial sectors by the State Council, to be fully implemented. During the 8th FYP, an annual environmental protection work plan was carried out in 1992, nationwide, based on the experience gained from the pilot work of compiling the annual environmental protection plan in some provinces and cities. Thus, in China, environmental protection was concretely incorporated into the national economic plan system and under the regulation and management of state planning.

3.2.4 Environmental Protection During the 9th FYP (1996–2000)

In March 1996, the trans-century implementation of sustainable development strategies and the goal of environmental protection were explicitly stipulated in the *Ninth*

FYP for National Economic and Social Development of the People's Republic of China and the Outline of the Long-Term Objectives for 2010, which were approved in the Fourth Session of the Eighth NPC. In July 1996, the State Council convened the Fourth National Conference on Environmental Protection, issued the *Decisions on Several Issues Concerning Environmental Protection* and ratified *the Ninth FYP for National Environmental Protection and the Long-Term Objectives for 2010*. Two significant measures, including Total Emission Cap and Trans-Century Green Project Plan, were implemented by the State to identify key areas for pollution governance, including "Three Rivers" (Huaihe River, Haihe River and Liaohe River), "Three Lakes" (Taihu Lake, Chaohu Lake and Dianchi Lake) and "Two-Control-Zone" (acid rain and sulfur dioxide control areas). Thus, local governments at all levels paid careful attention to the environmental protection plan and made great efforts to promote the implementation of the environmental protection plan, whereby the plan was concretely allocated to projects to significantly enhance the feasibility of the plan so that the compilation and implementation of the environmental protection plan became an essential link between environmental decision-making and management, and the main course of environmental protection work. The representative plan of this period was the water environment plan, for example, Three Rivers and Three Lakes. In addition, as the 21st century drew closer, the research and environmental protection plans for environmental problems of the 21st century became the new hotspots of the environmental protection plan.

3.2.5 Environmental Protection During the 10th FYP (2001–2005)

In early 2000, *the Compiling Technical Outlines of the Local Environmental Protection Plan during the Tenth FYP and the Long-Term Objectives for 2015* were formulated by the SEPA, the key features of which can be summarized as follows:

- to advocate persistent environmental protection as one of the national policies and sustainable development strategy;
- to improve environmental quality;
- to ensure national security;
- to protection human health;
- to pay equal attention to both pollution prevention and control and ecological protection;
- to implement three key measures, namely, a total emission cap on pollutants, zoning scheme for ecological protection and management, and Green Project Plan;
- to realize classified guidance based upon the watershed and regional environmental zoning.

On March 15, 2001, the 10th FYP was approved by the Fourth Session of the Ninth NPC to propose the following three key points: paying greater attention to population, resources, ecological and environmental issues; resolving issues concerning strategic resources (such as grain, water, and petroleum); and enhancing the implementation of a sustainable development strategy to reach a higher level. During the 10th FYP, the *Key Special Plans for Ecological Construction and Environmental Protection in the 10th FYP* was promulgated in August 2001, whereby the environmental protection plan was integrated into the key special plans for the first time.

In July 2001, the *National Environmental Protection Plan during the 10th FYP (exposure draft)*, compiled by the SEPA according to the requirements of the *Recommendations on the Formulation of the Tenth FYP for National Economic and Social Development* and the *Outline of the Tenth FYP for National Economic and Social Development*, was disseminated to various departments of the State Council as well as provinces, autonomous regions and municipalities for consultation. Some main issues were resolved coordinately. On December 26, 2001, the *National Environmental Protection Plan during the 10th FYP* was approved by Premier Rongji Zhu during the Prime Minister's Office Meeting. As pointed out in the *Rescriptum on the National Environmental Protection Plan during the 10th FYP*: *The National Environmental Protection Plan during the 10th FYP,* it "is an important reference for environmental work during the 10th FYP. Accordingly, all provinces, autonomous regions, and municipalities directly under the Central Government and the relevant departments under the State Council were expected to formulate their own specific implementation plan as soon as possible, and specifically incorporate key environmental protection projects into their own annual plans for national economic and social development, and implement the plans solidly."

It was reaffirmed in the 10th FYP for National Environmental Protection that environmental protection would be one of the national policies. Based upon the performance evaluation of the national environmental protection work carried out during the 9th FYP and the analysis on the current environmental situation, the overall objective of environmental protection during the 10th FYP was determined as follows: "By 2005, the status of environmental pollution will be alleviated; the trend of ecological degradation will be contained; urban-rural environmental quality, especially for large and medium-sized cities and key areas, will be improved; and the laws, policies and management system of environmental protection will be perfected to well adapt to the socialist market economic system."

There were a series of special plans meant as supplements to the *National Environmental Protection Plan during the 10th FYP*, including the *Plan for the Disintegration of Total Emission Cap on Major Pollutants during the 10th FYP* and the *Key Engineering Projects of the National Environmental Protection Plan during the 10th FYP*. In addition, the SEPA worked along with the SPC, SETC, and other relevant departments and regions to formulate three special plans (namely, the *National Ecological and Environmental Protection Plan during the 10th FYP*, the *Science and Technology Development Plan for National Environmental Protection during the 10th FYP* and the *National Environmental Monitoring Plan during the 10th FYP*) and the environmental protection plan for ten key areas designated in the "33211 Project" during the 10th FYP (including Three-River, Three-Lake, Two-Control-Zone, 1 City-Beijing, and 1 Ocean-Bohai Ocean). All these plans were the concrete contents and essential components of the *National Environmental Protection Plan during the 10th FYP*.

3.2.6 Environmental Protection During the 11th FYP (2006–2010)

Since 2006, the Chinese government has put much more efforts into environment protection. The extent and details of the content of environmental protection in the 11th FYP are unprecedented in putting forward the scientific outlook on development, setting up a special chapter of the specified provisions on the tasks and objectives of environmental protection, and instituting specific environmental protection guidance (such as improving resource utilization efficiency, strengthening pollution control, and promoting resource conservation).

As specified in the 11th FYP, the scientific outlook of development was to be comprehensively implemented by promptly transforming the models of economic growth from relying mainly on increasing investment in resource exploitation to improving the efficiency of resource utilization, on the basis of resource conservation. A regionalization concept was also proposed whereby the land was to be divided into four types of main functional zones (including optimized development, important development, restricted development, and prohibited development) with diverse developing themes and courses based upon their carrying capacity of resources and environment, where differentiated policies and rules applied (such as fiscal policy, investment policy, industrial policy, land policy, population management policy, performance evaluation policy, and so on) to guarantee the deployment of these zoning planning (Liu, Zhang, & Bi, 2012).

In order to build a new socialist countryside, specific objectives and requirements of strengthening rural environmental protection were promulgated, including the national soil pollution survey; comprehensive soil pollution management; prevention and control of pesticides, fertilizer and plastic sheeting and other non-point source pollution; pollution prevention and control of large-scale farms; enhancement of rural living garbage and sewage treatment; improvement of environmental sanitation and village appearance; and prevention of the transfer of industrial solid wastes, hazardous wastes, urban rubbish, and other pollutants to rural areas.

Moreover, the goal of building a resource-saving and environment-friendly society was stated in a special chapter to stipulate some important codes of conduct for building an environmentally friendly society. Additionally, the Circular Economy approach was declared for the first time to promote a balance between conservation and development while ensuring that conservation comes first, and to gradually establish a resource recycling system in various sectors, such as resource extraction, production and consumption, waste generation, consumption, and so forth, according to the principle of reduce, reuse, and recycle. In the aspect of resource management, natural resources protection and management were to be enhanced, and resource exploitation and utilization was to be limited, orderly, and compensated. By the end of the 11th FYP, energy intensity and major pollutants emissions were expected to be 10% less than that in 2005, and forest coverage was to be increased from 18.2% in 2005 to 20%. More importantly, the achievement of these goals were to serve as the indicators of performance evaluation for governments at all levels.

3.2.7 Environmental Protection During the 12th FYP (2011–2015)

The preliminary research on the *National Environmental Protection Plan during the 12th FYP* was initiated as early as November 2008 by the MEP to set up the principles, courses, targets, and focal points of the *National Environmental Protection Plan during the 12th FYP*, which was officially promulgated on December 15, 2011. As this was the critical period of building a well-off society (moderately prosperous society) in all aspects, several key tasks encompassing the theme of economic development transformation were outlined in the 12th FYP, such as transforming economic development models; adjusting economic structures; facilitating the coordination between first, second, and tertiary industries; strengthening the supporting foundation, benefiting people's livelihoods, and promoting a deeper reformation. Therefore, in due course, green development became the inevitable choice in the context of economic development transformation. Since the Sixth National Conference on Environmental Protection (April 2006), the relationship between environmental protection and economic development has changed drastically, due to the mutual need for green development.

Along with the advocacy and promotion of a scientific outlook on development, economic development transformation, and ecological civilization, environmental protection was smoothly integrated into the green transformation of the economy to further enhance its capacity building through the continuous expansion of its applications to diverse fields. In addition, in the context of the high incidence of environmental pollution incidents during the middle to later stages of industrialization, several strategic missions were promulgated for the first time, such as environmental risk prevention and control in key areas (with special emphasis on the safety and security requirements for nuclear and radiation environment), the improvement and perfection of the basic public service system of environmental protection (with a particular focus on the comprehensive improvement of the rural environment through the enhancement of rural environmental protection work).

A differentiated regional management system was initiated in diverse environmental policies. According to the *National Major Function Zoning Plan*, categorized guidance and regionalized management schemes were applied to four major function zones (such as optimized development zones, major development zones, limited development zones, and prohibited development zones) on the basis of the differences between various environmental functions, along with the application of the Ecological Redline to sensitive and fragile environmental zones. Furthermore, special environmental management requirements were proposed for four major function zones and four major economic zones (east, middle, west, and northeast China).

In order to strengthen the support toward policies and to promote the long-term mechanism of environmental protection, many measures were put forward, such as concretely implementing the Liability System of Environmental Protection Objectives, perfecting a comprehensive mechanism for decision-making, and so forth, where a strong emphasis was placed on the promotion and utilization of market mechanisms and the enhancement of scientific and industrial support. Furthermore, many schemes were announced to perfect the system of environmental policy, as shown in Box 3.1.

Box 3.1 Major supporting schemes of environmental policy
- *Special price policy for thermal power plant equipped with denitrification facilities.*
- *Special policy towards industries equipped with solid waste treatment facilities, wastewater treatment facilities, sludge treatment facilities, or desulfurization and denitrification facilities.*
- *To promote the emission trading market so as to improve the emission permit system.*
- *To establish the evaluation system of environmental performance credit for enterprises and the green finance rating system so as to promote the marketization of financing mechanism.*
- *To develop the ecological compensation mechanism for watershed and key ecological function zones.*
- *To enhance the capacity building of research on science and technology and their applications.*
- *To initiate the qualification system for environmental services industries and the permit system for the operation of environmental facilities.*

3.2.8 *Ecological and Environmental Protection During the 13th FYP (2016–2020)*

In November 2016, the Ecological and Environmental Protection Plan during the 13th FYP was promulgated by the State Council to present certain new features. First of all, for instance, the title was changed from "Environmental Protection" to

"Ecological and Environmental Protection" so as to realize the overall planning for ecological protection and environmental protection. Second, the notion of planning was modified to set the improvement of environmental quality as the core goal and evaluation criteria, where 12 binding indicators were proposed to highlight the systematic linkage between environmental quality improvement and the work of emission reduction, ecological protection, and environmental risk prevention and control. Third, targets and tasks of governance were allocated to regions, watersheds, cities and various control units, so as to implement refined and inventory management for environmental quality improvement. In water governance, the water environment throughout the country was divided into 1784 control units, where the objectives and improvement requirements were elaborated for 346 control units that exceeded the standards. In air governance, diverse objectives and improvement requirements were accordingly proposed for three large regions: Beijing-Tianjin-Hebei, the delta region of the Yangtze River, and the delta region of the Pearl River. Fourth, green development was designated as an important task for exploring solutions for ecological and environmental issues at the origin of development. In addition, dozens of important policy schemes and institutional reform were put forward to improve and ensure the implementation of the plan, in order to promote reform through its implementation (Wang et al., 2018), as shown in Table 3.3.

3.3 The Coordination and Integration of Multiple Plans

The institutional structure of the Chinese government is a very sophisticated and massive matrix due to vertically (from central government to local governments) and horizontally (various departments within the same government level) intertwined power structures (Lieberthal, 2004). Therefore, planning has become one of the key measures of macro control and management for the Chinese government. Normally, governments and departments at all levels strive for resources and power through the compiling of plans. Based on the National Economic and Social Development Plan, the Chinese planning system evolved and formed spontaneously through long-term development. There are massive plans with diverse categories, but they lack coordination (Wu, Song, Lin, & He, 2018). For example, according to the *Law of the People's Republic of China on Environmental Impact Assessment*, environmental impact assessment is required for 14 different types of plans, including land utilization, watersheds, marine areas, regions and 10 special plans (Zhu, Jing, & Chang, 2005). Moreover, during the 10th FYP, more than 7300 plans, including development plans, key special plans, and sector plans, were compiled by provincial, city, and county governments. Thus, on August 26, 2014, the *Notice Concerning the Pilot Work of the Coordination and Integration of Multiple Plans at City and County Level* was jointly disseminated by the NDRC, MLR, MEP, and MHURD to propose the pilot program of spatial planning reform at 28 cities and counties, so as to promote the coordination and integration of multiple plans, including economic and social development plans, urban and rural development plans, land utilization plans, and

Table 3.3 Major targets of ecological and environmental protection during the 13th FYP

Indicators		2015	2020	Cumulative[a]	Nature
Eco-environmental quality					
1. Air quality	Percent of days with good air quality in cities at prefecture-level and above[b] (%)	76.7	>80	–	Binding
	Reduction of fine particle concentration in cities at prefecture-level or above failing to meet the standard (%)	–	–	[18]	Binding
	Decline in the percent of days with heavy pollution or even worse in cities at prefecture-level or above (%)	–	–	[25]	Expected
2. Water quality	Percent of surface water with quality at or better than Grade III[c] (%)	66	>70	–	Binding
	Percent of surface water with quality worse than Grade V (%)	9.7	<5	–	Binding
	Percent of major rivers and lakes attaining water quality standards (%)	70.8	>80		Expected
	Percent of ground water with very poor quality (%)	15.7[d]	≈15	–	Expected
	Percent of coastal waters with excellent and good water quality (I, II) (%)	70.5	≈70	–	Expected
3. Soil quality	Safe utilization rate of contaminated farm land (%)	70.6	≈90	–	Binding
	Safe utilization rate of contaminated fields (%)	–	>90	–	Binding
4. Ecological conditions	Forest coverage (%)	21.66	23.04	[1.38]	Binding
	Forest stock volume/10^8 m^3	151	165	[14]	Binding
	Stock wetland/667 km^2	–	≥8	–	Expected
	Vegetation coverage of grassland (%)	54	56		Expected
	Environmental condition index of counties in areas with key ecological functions	60.4	>60.4	–	Expected

(continued)

3.3 The Coordination and Integration of Multiple Plans

Table 3.3 (continued)

Indicators		2015	2020	Cumulative[a]	Nature
Emission caps					
5. Reduction of major pollutants discharge (%)	Chemical oxygen demand (COD)	–	–	[10]	Binding
	Ammonia nitrogen	–	–	[10]	Binding
	Sulfur dioxide (SO_2)	–	–	[15]	Binding
	Nitrogen oxides (NO_x)	–	–	[15]	Binding
6. Reduction of regional pollutants discharge (%)	Volatile organic compounds (VOCs) of key industries in key regions[e]	–	–	[10]	Expected
	Total nitrogen (TN) in key regions[f]	–	–	[10]	Expected
	Total phosphorous (TP) in key regions[g]	–		[10]	
Ecological conservation and restoration					
7. Protection rate of wildlife under national priority protection (%)		–	>95	–	Expected
8. Natural shoreline retention rate (%)		–	⩾35	–	Expected
9. Newly protected land under desertification control/10^4 km^2		–	–	[10]	Expected
10. Newly protected land under water and soil erosion control/10^4 km^2		–	–	[27]	Expected

[a]Five-year cumulative number in brackets [ ]
[b]Air quality assessment covers 338 cities nationwide (including prefecture-level, league-level and some county-level cities under provincial jurisdiction, excluding Sansha and Danzhou)
[c]Water environmental quality assessment covers surface water sections under national monitoring program which increased from 972 (during the 12th FYP period) to 1,940
[d]Data of 2013
[e]More than 10% of total VOCs emission would be cut through strengthened control over its emission in key industries and key regions
[f]Total TN control covers 56 coastal cities and 29 eutrophic lakes and reservoirs
[g]Total TP control covers units with excessive TP emissions and related up stream areas

ecological and environmental protection plans. The purpose of the coordination and integration of multiple plans is to explore the appropriate mechanism to generally incorporate the ecological and environmental protection plans into the national economic and social development plans, urban and rural development plans, and land utilization plans, as shown in Fig. 3.1. Hence, it is an important tool and a mean of facilitating the integration of ecological and environmental protection into comprehensive decisions and the promotion of sustainable development and ecological civilization.

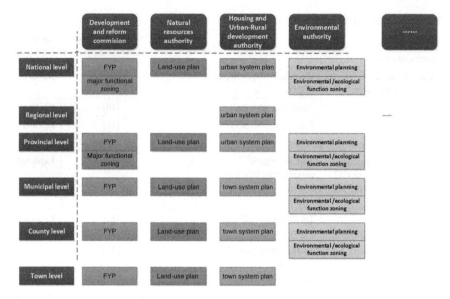

Fig. 3.1 The scope of plans for coordination and integration

References

Jin, M. (2014). *Environmental management*. Beijing: China Environment Press.
Lieberthal, K. (2004). Governing China.
Liu, L., Zhang, B., & Bi, J. (2012). Reforming China's multi-level environmental governance: Lessons from the 11th FYP. *Environmental Science & Policy, 21,* 106–111. https://doi.org/10.1016/j.envsci.2012.05.001. [2019-04-11].
Wang, J., Wan, J., Wang, Q., Su, J., Yang, L., & Xiao, Y. (2018). The development of China's ecological and environmental planning in forty years of reform and opening-up. *Chinese Journal of Environmental Management, 10*(6), 5–18.
Wu, J., Song, Y., Lin, J., & He, Q. (2018). Tackling the uncertainty of spatial regulations in China: An institutional analysis of the "multi-plan combination". *Habitat International, 78,* 1–12. https://doi.org/10.1016/j.habitatint.2018.07.002. [2019-04-12].
Zhu, T., Jing, W. U., & Chang, I. S. (2005). Requirements for strategic environmental assessment in China. *Journal of Environmental Assessment Policy & Management, 7*(01), 81–97.

Chapter 4
Environmental Impact Assessment

In 1969, *the National Environmental Policy Act* was firstly promulgated in the USA, where environmental impact assessment (EIA) was designated as an important and mandatory requirement for environmental management (Section 102 [42USC§4332]). Later on, many countries integrated EIA into their environmental management systems as an essential requirement, worldwide. Therefore, EIA, as a vital means for both environmental management and sustainable development, has become the precautionary measure of environmental protection. In addition, EIA is also an important tool to play an essential role in coordinating and balancing the relationship between economic development and environmental protection.

4.1 The Concept and Principles of EIA

The major function of EIA is to systematically diagnose, analyze, predict and appraise the environmental impacts or consequences would be caused by proposed development activities, such as projects, plans and policies, through careful and thorough consideration of all relevant environmental information, during the decision-making process. In other words, the purpose of EIA is to integrate the goals of environmental protection with social and economic development to prevent and mitigate adverse impacts of the proposed activities, as much as possible. Therefore, the objectives of EIA can be summarized as:

- to incorporate environmental considerations into developing proposals and decision making;
- to predict, prevent, mitigate or counteract adverse effects from developing proposals;
- to preserve and maintain ecosystem and natural processes;
- to ensure rational utilization and management of natural resources;
- to promote sustainable development.

In order to accomplish these objectives, the practice of EIA should conform to the following principles, including purposeful, rigorous, practical, efficient, adaptive, focusing, participative, transparent, systematic, integrated, credible, and cost-effective (Joseph, Gunton, & Rutherford, 2015; Senecal, Goldsmith, Conover, Sadler, & Brown, 1999).

4.2 The Development of EIA Globally

The concept of EIA was first proposed in the International Academic Conference on Environmental Quality Assessment, held in Canada in 1964. In the USA, EIA was explicitly stipulated in the "*National Environmental Policy Act*" and enacted in 1970. Ever since, the EIA system was adopted in many countries, including Sweden (1970), New Zealand (1973), Canada (1973), Australia (1974), Malaysia (1974), Germany (1976), India (1978), Philippine (1979), Thailand (1979), China (1979), Indonesia (1979), Sri Lanka (1979), and so on, where the EIA system was explicitly stipulated in relevant legal provisions, such as the Environmental Protection Act (1988) of Canada, Environmental Law (1993) and Environmental Impact Assessment Law (1997) of Japan, General Provisions of Environmental Protection Law of Netherland and Environmental Impact Assessment Law (1990) of Germany (Glasson, Therivel, & Chadwick, 2011).

The first EIA meeting was held in 1974 by the United Nations Environment Programme (the UNEP) and Canada. EIA requirements are then gradually enshrined in many international agreements and international law (Craik, 2008). Into the 1990s, the EIA commitments were generally integrated into many environmental treaties, including the Rio Declaration on Environment and Development, the Agenda 21, the Convention on Biological Diversity (the CBD), and the United Nations Framework Convention on Climate Change (the UNFCCC), and so on. Detailed requirements for the conduct of EIAs were also instituted in many binding treaties, such as the Convention on Environmental Impact Assessment in a Transboundary Context (the Espoo Convention) developed by the United Nations Economic Commission for Europe (the UNECE), and the Protocol on Environmental Protection to the Antarctic Treaty (the Antarctic Protocol). Some international organizations incorporated the EIA norms into their own requirements. The most prominent set of EIA requirements adopted by international organizations are those of the World Bank and the Asian Development Bank. Henceforth, the EIA system gradually became an important part of environmental management, worldwide, and was continuously developing and improving.

4.3 The Development of EIA in China

Since the EIA concept introduced into China in 1970s, an EIA system was initiated, evolved and developed into a legal system through the continuous legislative procedures and improvement, where the Central Government was taking the leading role. The EIA Law (2002) was promulgated in October 2002 to mark a milestone of the legal and systematic development of EIA in China. EIA is not only a technique but also a legal statute for environmental management. Generally, the development of the EIA system in China can be divided into several phases, as illustrated in the followings.

4.3.1 Initial Phase (1973–1978)

In 1972, the Chinese delegation participated in the United Nations Conference on Human Environment held in Stockholm. In response to the call of the Conference, the Preparatory Office for the LGEP was established under the State Council to be in charge of environmental protection work, coordinated by the SPC. In August 1973, the First National Conference on Environmental Protection was held in Beijing to promulgate *Several Regulations Concerning Protecting and Improving Environment (on Trial Draft)* where the key guiding principles of environmental protection in China were outlined, including overall planning, rational layout, comprehensive utilizations, converting disadvantage into advantage, public participation, mobilizing everyone, protecting the environment, and benefiting all. In addition, environmental protection institutions were then set up within local governments and relevant departments to supervise and examine environmental protection work, as instructed by the State Council. Since the First National Conference on Environmental Protection, the concept of EIA was formerly introduced into China. And, EIA was disseminated and advocated by experts and scholars from higher educational institutions and scientific research institutes during academic conferences and research papers.

4.3.2 Early Implementation Phase (1979–1989)

In April 1979, EIA was re-endorsed as one guiding policy by the Environment Protection Leading Group in *the Progress Report of the National Environmental Protection Work Meeting*. And the first EIA for construction project (Project EIA) was carried out for the Yongping Copper Mine, in Jiangxi Province. In May 1979, EIA was specifically regulated as the requirement for construction projects in *the Notice Concerning Well Preparation for the Preliminary Work of Construction Project* by the SPC and the State Construction Commission (the SCC). In September 1979, the EP

Law (on Trial) was promulgated to explicitly regulate the requirement of EIA in Article 6, as shown in Box 4.1.

> **Box 4.1 Article 6 of the EP Law (on Trial)**
> - *All enterprises and institutions must pay adequate attention to the prevention of pollution and damage to the environment when selecting their sites, and in designing, construction and production. In planning new construction, reconstruction and extension projects, the units concerned must submit an environmental impact statement to the environmental protection department and other relevant departments for examination and approval before the designing is started. Installations for the prevention of pollution and other hazards to the public must be designed, built and put into operation simultaneously with the main project. Discharge of all kinds of harmful substances must be in compliance with the standards set by the State.*

Later on, several laws, regulations, ordinances, and administrative rules regarding to environmental protection were consecutively promulgated to further regulate the EIA system, to clearly formalize the content, scope and procedures of EIA, and to constantly improve the methodologies and techniques of EIA. In May 1981, *the Measures of Management for Environmental Protection of Infrastructure Construction Project* was promulgated jointly by the SPC, the State Construction Commission (the SCC), the State Economy Commission (the SEC), and the LGEP, where the EIA system was definitely integrated into the review and approval procedures for construction projects as the mandate requirement. In 1986, the scope, content, review and approval procedures, and the format of the EIA Report for Project EIA were clearly defined in the Measures of Management for Environmental Protection of Construction Project, promulgated jointly by the SPC, the SEC and the Environmental Protection Commission (the EPC) of the State Council, to facilitate the effective implementation of EIA system. In the same year, the Measures of Management for the Certificate System of Environmental Impact Assessment of Construction Project (on Trial) was promulgated by the NEPA to initiate a license system of EIA consultation agencies. In addition, in order to accentuate the importance of Project EIA, Project EIA was explicitly stipulated as a legal requirement in many articles of other relevant laws, for example, Articles 6, 9, and 10 of *the Law of the People's Republic of China* on *Marine Environment Protection* (1982), Article 13 of *the Law of the People's Republic of China on Water Pollution Prevention and Control* (1984), Article 9 of *the Law of the People's Republic of China on Air Pollution Prevention and Control* (1987), Article 12 of *the Law of the People's Republic of China on Wildlife Protection* (1988), and Article 15 of *the Ordinance of the People's Republic of China on Environmental Noise Prevention and Control* (1989).

After ten years of practice and improvement, the EP Law (on Trial), promulgated in 1979, was formally finalized to become the EP Law (1989) promulgated on December 26, 1989. The objectives, missions, work principles, review and approval procedures, and implementation time of EIA and the relationship between EIA and construction process were all explicitly stipulated in Article 13 of the EP Law (1989). In the EP Law (1989), it is of great importance to reaffirm the legal status of Project EIA and to provide the concrete foundation for the implementation of Project EIA.

4.3.3 Improvement Phase (1990–2002)

Since 1990s, the open-up and reform of social development and economic system in China have stimulated tremendous transformation on political, social, and economic system. Hence, from the EP Law (1989) to the EIA Law (2002), the development of EIA system has been significantly improved and strengthened through solid construction and fast expansion of objectives, scope, content, methodologies, and professional practitioners of EIA. Furthermore, regional environmental impact assessment (REIA) was then initiated. In addition to environmental management, EIA for ecological construction project was also reinforced to concentrate on ecological conservation, and pollution prevention and control, concurrently. Through the implementation of the project supported by the World Bank (the WB) and the International Finance Corporation (the IFC), public participation in Project EIA was introduced in China. The scope and approach of public participation were gradually expanded and improved.

Since 1994, several technical guidelines for EIA, were promulgated and enacted, including EIA guideline-general requirements, EIA guidelines for environmental elements (such as surface water, atmospheric environment, noise, and etc.), and EIA guidelines for specific industrial sector, as displayed in Table 4.1.

In August 1996, *the Decisions on Several Issues Concerning Environmental Protection* was promulgated by the State Council during the Fourth National Conference on Environmental Protection to strengthen the review and approval procedures for construction projects, to implement total emission cap, to add the requirements of cleaner production and public participation, to enhance ecological EIA, and to further expand the depth and breadth of EIA. In addition, pilot studies on EIA follow-up were conducted on several trial projects to accumulate experiences.

In November 1998, *the Ordinances of Management for Environmental Protection of Construction Project*, the first administrative statute concerning environmental management for construction projects in China, was promulgated by the State Council. In March 1999, based upon this administrative statute, *the Measures of Management for the Qualification and Certificate System of Environmental Impact Assessment of Construction Project* was promulgated by the SEPA to prescribe the qualification of the institution to perform EIA. In April 1999, a list of classification management was publicized through the promulgation of *the Catalogue of Classification Management for Environmental Protection of Construction Projects (on Trial)*

Table 4.1 List of various EIA guidelines in China (updated in January 2019)

Title	Latest version	Previous version
EIA guideline-general requirements		
Technical Guidelines for Environmental Impact Assessment for Construction Project—General Program	HJ 2.1—2016	HJ 2.1—2011 HJ/T 2.1—1993
Guidelines for Technical Review of Environment Impact Assessment for Construction Project	HJ 616—2011	
EIA guideline-sector specific		
Technical Guidelines for Environmental Risk Assessment on Tailings Pond	HJ 740—2015	
Technical Guideline for Environmental Impact Assessment for Iron and Steel Construction Project	HJ 708—2014	
Technical Guidelines for Environmental Impact Assessment for Electric Power Transmission and Distribution Project	HJ 24—2014	HJ/T 24—1998
Technical Guidelines for Environmental Impact Assessment for Coal Development Project	HJ 619—2011	
Technical Guidelines for Environmental Impact Assessment for Pharmaceutical Construction Project	HJ 611—2011	
Technical Guidelines for Environmental Impact Assessment for Pesticide Construction Project	HJ 582—2010	
Technical Guidelines for Environment Impact Assessment on Urban Rail Transit	HJ 453—2018	HJ 453—2008
Technical Guideline for Environmental Impact Assessment for Terrestrial Petroleum and Natural Gas Development Construction Project	HJ/T 349—2007	
Technical Guidelines for Environmental Risk Assessment on Projects	HJ 169—2018	HJ/T 169—2004
Technical Guidelines for Regional Environmental Impact Assessment on Development Area	HJ/T 131—2003	
Technical Guidelines for Environmental Impact Assessment for Water Conservancy and Hydropower Project	HJ/T 88—2003	
Technical Guidelines for Environmental Impact Assessment for Petrochemical Construction Project	HJ/T 89—2003	
Technical Guidelines for Environmental Impact Assessment for Civil Airport Construction Project	HJ/T 87—2002	
Technical Regulations on Environmental Impact Assessment of Electromagnetic Radiation Produced by 500 kV Ultrahigh Voltage Transmission and Transfer Power Engineering	HJ/T 24—1998	

(continued)

4.3 The Development of EIA in China

Table 4.1 (continued)

Title	Latest version	Previous version
Guidelines for Radioactive Environment Protection and Management/the Methods and Standards of Environmental Impact Assessment on Electromagnetic Radiation	HJ/T 10.3—1996	
EIA guidelines-environmental elements		
Technical Guidelines for Environmental Impact Assessment on Soil Environment	HJ 964—2018	
Technical Guidelines for Environmental Impact Assessment on Groundwater Environment	HJ 610—2016	HJ 610—2011
Technical Guidelines for Environmental Impact Assessment on Ecological Impacts	HJ 19—2011	HJ 19—1997
Technical Guidelines for Environmental Impact Assessment on Noise Environment	HJ 2.4—2009	HJ/T 2.4—1995
Technical Guidelines for Environmental Impact Assessment on Atmospheric Environment	HJ 2.2—2018	HJ 2.2—2008 HJ/T 2.2—93
Baseline of Environmental Quality Risk Assessment for Soil at Manufacturing Facilities	HJ/T 25—1999	
Technical Guidelines for Environmental Impact Assessment on Surface Water Environment	HJ 2.3—2018	HJ/T 2.3—93 (1993)
Plan EIA guidelines		
Technical Guidelines for Plan Environmental Impact Assessment—General Principles	HJ 130—2014	HJ/T 130—2003
Technical Guidelines for Plan Environmental Impact Assessment—The Master Plan for Coal Industry Mining Area	HJ 463—2009	

by the SEPA. The promulgation of the EIA Law (2002) in October 2002 and entering into force in September 2003 marked the end of the improvement phase. During this improvement phase, the SEPA has strengthened the qualification management for professionals of Project EIA institutions, and trained EIA practitioners to employ qualification system for Project EIA.

4.3.4 Breakthrough Phase (2003–2014)

As a result of the promulgation of the EIA Law (2002), the objectives of EIA were extended from construction projects to sector plans of government. The SEPA established the basic database of EIA and issued *the Technical Guidelines for Environmental Impact Assessment for Plans (on Trial)* in October 2003. In addition, as approval by the State Council, the list of plans subject to EIA was constituted by the SEPA, jointly with several relevant departments, as well as the promulgation of *the Measures of Management for the Reviewing Experts Database of Environmental Impact Assessment* (in August 2003) and *the Measures of Reviewing Environmental Impact Assessment Report of Specific Plan* (in October 2003) to gradually establish the procedures of EIA review, systematically.

In February 2004, the Ministry of Personnel, jointly with the SEPA, established the certification system for professional EIA engineers to put more stringent requirements on professional EIA practitioners. In August 2005, the *Measures of Management for the Qualification and Certificate System of Environmental Impact Assessment of Construction Project* was revised by the SEPA to impose higher liability for professional EIA practitioners, and to build an integrated management system of professional EIA institutes and individuals, along with professional EIA license system.

In February 2006, the *Interim Measures of Public Participation in Environmental Impact Assessment*, promulgated by the SEPA, was the first document regulating public participation in environmental protection in China. It clearly defined the rights of public to participate in EIA, and explicitly prescribed the specific scope, procedures, channels and time limit of public participation, which is beneficial to assure public the "Right to Know" of the environmental information.

4.3.5 Reform (2015–Date)

During the past 40 years, the procedures, process and practice of EIA was continuously evolving with the rapid development of social, economic and political context. There are still some substantial shortcomings with the current EIA system which limit its effectiveness as a tool for promoting sustainable development. Weak enforcement, insufficient public participation has blighted the EIA system. Moreover, EIA for plan (PEIA) and strategic environmental assessment (SEA) are usually excluded from the planning process, given that, for most of the non-environmental departments, PEIA and SEA are generally regarded as the external interference in executive power and eventually rejected. Against that backdrop, the MEP reviewed the EIA system and proposed some major modifications in *the Implementation Procedures of Environmental Impact Assessment Reform during the 13th FYP*, promulgated in July 2016.

① To redefine the orientation, relationship and function of SEA, PEIA and Project EIA. The task of SEA is to coordinate environmental issues arising from regional or cross-regional development, to designate ecological red line areas, and to provide the basis for "the coordination and integration of multiple plans" and PEIA. PEIA focuses on optimizing the layout, scale and structure of industrial development, drawing up the inventory of environmental management and control, and guiding the environmental compliance of developing projects. The Project EIA focuses on fulfilling the requirements for environmental management, optimizing the measures of environmental protection, strengthening environmental risk prevention and control, and coordinating with the environmental permits system.

② To simplify the review and approval for Project EIA. Since 2013, the authority to review and approve EIA documents for 57 types of construction projects has been gradually transferred from the MEP to provincial EPBs. Then, this authority was further empowered to cities, counties and districts by provincial EPBs, according to practical local conditions. Since 2016, the review and approval for the Environmental Impact Registration Form (the EIRF) for Construction Project has been cancelled and replaced by reporting and archiving process. The scope of projects required to compile the Environmental Impact Assessment Report (the EIA Report) was reduced through the revision of *the Catalogue of Classification Management for Environmental Impact Assessment of Construction Projects* (the EIA Catalogue, first published in 1999; revised in 2002, 2008, 2015, 2017 and 2018, respectively).

③ To repeal the certificate system for EIA consultancies. The certificate system for EIA consultancies was introduced in 1986 by the NEPA. The system was originally designed to ensure the quality of EIA working task. For more than 30 years, the system has been playing an important role in ensuring the quality of EIA documents and strengthening the effectiveness of the EIA system. By the end of 2017, there were 920 certified EIA consultancies with 10,962 registered EIA professional engineers. In the context of the government's administrative and management reform, in order to promote market competition in the EIA consulting service and to revoke the limitations on market access, the certificate system for EIA consultancies was repealed in the amendment of the EIA law in 2018.

④ To strengthen the enforcement of the EIA Law. Key features include to strengthen progress monitoring and post-supervision on EIA, to perfection the liability mechanism for any EIA violation, to enhance the ability of review and approval, and supervision on EIA for the competent authority for ecology and environment of local governments, to reinforce public participation and information disclosure in EIA, and to effectively coordinate the EIA system, the Three Synchronizations system and the Emission Permit system.

4.4 The Legal Framework of EIA in China

In China, the development of EIA system was gradually improved and reinforced along with prompt development of the legal system and administrative management system of environmental protection. Since promulgation of *the Law of the People's Republic of China on Environmental Protection (on Trial)* in September 1979, Project EIA was stipulated as a legal system. Many regulations, statutes, ordinances, and directives were issued consecutively to standardize the EIA system, to elucidate the content, scope and procedures of EIA, and to refine the methodologies and technologies of EIA. As the EP Law (1989) was finalized and formally promulgated, the objectives, missions, principles, review and approval procedures and time frame of EIA were fundamentally prescribed in Article 13. Project EIA and EIA for Plan (PEIA) were precisely defined as the legal requirement in the EIA Law (2002) to provide the legal basis for the deployment of Strategic Environmental Assessment (SEA), and to promote the legalization and institutionalization of SEA in China. The laws and regulations with EIA requirements are listed in Table 4.2.

Table 4.2 Legal framework on EIA in China (updated in January 2019)

Title	Evolvement
Specific law on EIA	
Environmental Impact Assessment Law of the PRC	Enacted in 2002, amended in 2016, 2018
Specific law on environmental protection with EIA requirements	
Environmental Protection Law of the PRC	Enacted in 1979, revised in 1989, 2014
Air Pollution Prevention and Control Law of the PRC	Enacted in 1987, amended in 1995, 2000, 2015, 2018
Water Pollution Prevention and Control Law of the PRC	Enacted in 1984, revised in 2008, amended in 1996, 2017
Noise Pollution Prevention and Control Law of the PRC	Enacted in 1996
Solid Wastes Pollution Prevention and Control Law of the PRC	Enacted in 1995, revised in 2004, amended in 2013, 2015, 2016
Soil Pollution Prevention and Control Law of the PRC	Enacted in 2018
Marine Environment Protection Law of the PRC	Enacted in 1982, revised in 1999, amended in 2013, 2016, 2017
Radioactive Pollution Prevention and Control Law of the PRC	Enacted in 2003
Cleaner Production Promotion Law of the PRC	Enacted in 2002, amended in 2012
Circular Economy Promotion Law of the PRC	Enacted in 2008, amended in 2018
Water Law of the PRC	Enacted in 1988, revised in 2002, amended in 2009, 2016
Water and Soil Conservation Law of the PRC	Enacted in 1991, amended in 2010, revised in 2010
Wild Animal Conservation Law of the PRC	Enacted in 1988, amended in 2004, 2009, revised in 2016
Administrative regulations on EIA	
Regulations on the Administration of Construction Project Environmental Protection	Enacted in 1998, amended in 2017
Regulation on Environmental Impact Assessment for Plan	Enacted in 2009

4.5 EIA for Project (Project EIA)

What is project? There is no specific definition of "project" in our legislation. In *the Management Measures of Environmental Protection for Construction Project* (1986), the scope of project was clearly defined to encompass the following categories of construction projects, including industry, transportation, water conservancy, agriculture, forestry, business, education, tourism, municipal engineering sectors, and so forth. In addition, EIA is not only applied to new projects, but also to projects of reconstruction, modification and expansion. The guidelines, referred to the EIA Catalogue, were published by the administrative authorities to help proponents and EIA regulatory departments determine whether the EIA should be carried out for a project and what type of EIA should be performed, according to the location, industry sector, project scale, investments, types and amounts of pollutants, and the significance of environmental impacts. As specifically stipulated in the EIA Catalogue, the sensitivity of geographical location of a construction project is an important basis to determine which category the construction project should apply to. The sensitive areas were clearly defined in the EIA Catalogue, as shown in Box 4.2.

> **Box 4.2 Sensitive area defined in the EIA Catalogue**
> *Environmental sensitive area refers to various protection areas designated by laws and areas are particularly sensitive to environmental impacts of the construction project, mainly including areas within the Ecological Protection Redline, and the following areas:*
> - *nature reserves, scenic and historic interest areas, sites of world cultural and natural heritage, special marine protection areas and reserves for drinking water source;*
> - *the basic farmland protection areas, grasslands, forest parks, geological parks, important wetlands, natural forest, wildlife habitats, key sheltered grounds for wild plants breeding, natural spawning grounds, feeding grounds, wintering grounds and migration channels for important aquatic organisms, natural fishery, key areas for prevention and control of soil and water erosion, protection areas for desertification, closed and semi-closed sea area;*
> - *the functional areas of residence, medical and health care, cultural education, scientific research and administrative offices, and culture relic protection sites.*

4.5.1 Categorization for Project EIA

As stipulated in Article 16 of the EIA Law, Project EIA should be categorized according to the degree of environmental impact of construction project, as described in the following:

① where the potential impact is "significant", the developer shall prepare an EIA Report, in which a comprehensive EIA should be carried out;
② where the potential impact is "light", the developer shall prepare an EIA Form, in which detailed analyses or special EIA for specific environmental elements should be carried out;
③ where the potential impact is "too little" to require an assessment of them, the developer shall file an EIRF.

4.5.2 Process for Project EIA

The process for Project EIA is generally divided into three stages, as illustrated in Fig. 4.1. During the first stage, the administration authority shall determine whether an EIA Report, an EIA Form or an EIRF is required. If an EIA Report is required, an EIA work program (scoping report) should be produced based on environmental baseline investigation and preliminary engineering analysis of the proposed project. In the second stage, the changes in key indicators for those environmental components identified in the work program shall be predicted and evaluated. For the third stage, an EIA Report or an EIA Form is produced to describe the overall EIA process in detail.

4.5.3 Documentation for Project EIA

According to the significance of environmental impacts, the EIA documents for Project EIA in comprised of three categories, such as an EIA Report, an EIA Form, and an EIRF. In order to ensure the quality of EIA, to supervise the developer to fulfill the obligations in EIA, and to regulate the compilation of the EIA documents, the content and format of an EIA Report, an EIA Form, and an EIRF were explicitly regulated in the EIA Law, as shown in Box 4.3, and the formats for an EIA Form and an EIRF were also issued by the SEPA.

4.5 EIA for Project (Project EIA)

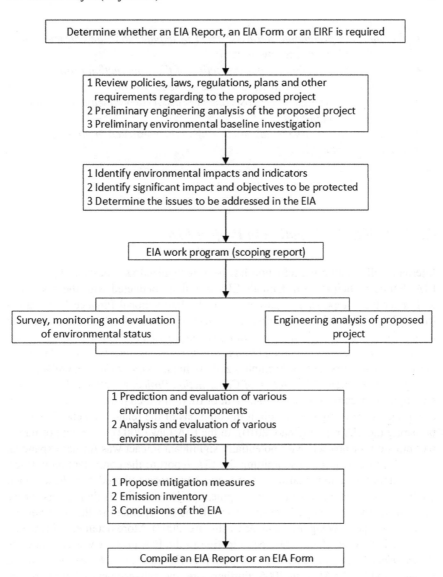

Fig. 4.1 The process for project EIA. *Source* Technical guideline for environmental impact assessment of construction project—general program (HJ 2.1—2016)

Box 4.3 Contents of an EIA report regulated in the EIA Law

The EIA Report of a construction project should include the followings:
- a brief introduction to the construct project;
- the current surrounding environment of the construction project;

- *the analyses, prediction and assessment of impacts that the construction project may have on the environment;*
- *the measures of environmental protection of the construction project, and technical and economic demonstrations of the measures;*
- *the analyses of the economic costs and benefits of environmental impacts of the construction project;*
- *a proposal for environmental monitoring of the construction project;*
- *conclusions of the EIA.*

4.5.4 Public Participation in Project EIA

Internationally, public participation has been recognized as a basic component of EIA. Public participation in China's EIA was first conducted with the assistance and under the compulsory requirements of the International Finance Corporation (the IFC) who funded the development projects (Zhang et al., 2012). As described in *the Notice on Strengthening the Management for EIA for Construction Projects loaned by International Financial Organizations* issued by the NEPA in 1993, public participation was first expressly emphasized in China. As *the Ordinances of Management for Environmental Protection of Construction Project* adopted in 1998, public participation became a formal component of EIA, in which developers of construction projects were required to solicit the opinions of the related stakeholders when preparing the EIA report (Zhao, 2010). In the EIA Law (2002), the extent of public participation for projects with potentially significant impact was further expanded. For such projects, prior to submitting the EIA report to the competent authorities, a consultation meeting, hearing, or an equivalent process should be held to solicit public opinions on the proposed development projects and its likely impacts. These opinions and the developer's response should be incorporated in the final version of the EIA report (Wang, Morgan, & Cashmore, 2003). More detailed guidance on who and how to conduct public participation in the EIA process was stipulated in *the Interim Measures of Public Participation in Environmental Impact Assessment* promulgated by the MEP in 2006. Furthermore, the enhancement of participatory rights during the EIA process was further ensured through the promulgation of *the Ordinances of the People's Republic of China on Government Information Disclosure* and *the Measures of Environmental Information Disclosure (on Trial)* in 2007 (Li, Thomas Ng, & Skitmore, 2012).

Though Project EIA has been implemented in China for more than 30 years, the public do not have any access to full text of the EIA report, until 2014. The public did not have sufficient information regarding the project while were invited or requested to fill out the questionnaire for EIA public survey. Since *the Interim Measures of Public Participation in Environmental Impact Assessment* came into force in 2006,

4.5 EIA for Project (Project EIA)

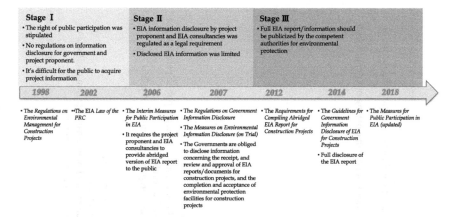

Fig. 4.2 Evolution of public participation and information disclosure in EIA in China

the status was improved, to some extent. The public could then have the access to abridged version of the EIA report provided by developer and EIA consultancy. From 2006 to 2013, the brief information and abridged version of the EIA report were just publicized on websites of EIA consultancy, which is very difficult for the public to notice. Even if the information was acquired, it would be too complicated to understand and too short for the public to deploy any concrete activities within 10 days, as the time window for soliciting public opinions. As the promulgation of *the Guidelines for Disclosure of Government Information on EIA for Construction Project* in 2013, the public can have access to full text of the EIA report easily from the website of EPB (Wu, Chang, Yilihamu, & Zhou, 2017). In 2018, *the Measures of Public Participation in Environmental Impact Assessment* were updated to further strengthen the enforcement of public participation in EIA. The evolution of public participation and information disclosure in EIA in China is illustrated in Fig. 4.2.

As specified in the EIA Law (2002), the requirements of public participation should only be applied to the construction project where the EIA report is prepared (see categorization for Project EIA). The public can participate in Project EIA at various stages. The key stages and participation opportunities in Project EIA process are displayed in Table 4.3.

4.5.5 Review and Decision on the EIA Documents

There are three levels of review and approval for EIA documents, the MEE; provincial (including autonomous region and municipality directly under the central government) EPBs; and municipal EPBs. Since 2013, as the review and approval process been continuously simplified according to practical experience, the authority of review and approval for EIA documents has been gradually transferred from the

Table 4.3 Key EIA stages and the requirements of public participation

Stages	Requirements
The initiation of the EIA process	The developer shall disclose the following information on the developer's website, websites of local public media, or local government websites (hereinafter referred to as the "Internet site"): 1. basic information of the construction project, such as name, site and activities involved in carrying out the proposed project; 2. name and contact information of the developer; 3. name of the EIA consultancy; 4. where to download the public opinion form; 5. how to submit the public opinion form
When the EIA report is drafted	The developer shall disclose the following information: 1. where to access the draft EIA report; 2. to whom to solicit the opinions; 3. where to download the public opinion form; 4. how to submit the written comments; 5. when to submit the written comments The period of soliciting public opinions shall not be less than 10 working days If many public opinions toward the environmental impacts are negative or doubting, more in-depth public participation activities, such as public meetings, hearings and workshops, shall be held by the developer The developer shall carefully analyze all opinions and comments acquired, and prepare a statement of public participation to describe the following information: 1. the process, scope and activities of public participation; 2. how the opinions and comments were collected, organized, analyzed and summarized; 3. an explanation of the developer's response to the public's opinions and comments The developer shall make both the full version of the EIA report and the statement of public participation available on the website before submitting to the competent authority for ecology and environment for approval

(continued)

4.5 EIA for Project (Project EIA)

Table 4.3 (continued)

Stages	Requirements
When the competent authority for ecology and environment received the EIA report and the statement of public participation	Once the competent authority for ecology and environment received the EIA report, the following information should be posted on the website: 1. the full version of the EIA report; 2. the statement of public participation; 3. how to submit the comments The period for information disclosure shall not be less than 10 working days
Before the decision on the EIA report is made by the competent authority for ecology and environment	The competent authority for ecology and environment shall disclose the following information on website before making the decision on the EIA report: 1. name and site of the construction project; 2. name of the developer; 3. name of the EIA consultancy; 4. activities involved in carrying out the proposed project, major environmental impacts, proposed countermeasures; 5. activities of public participation undertaken by the developer; 6. how to submit the opinions and comments The period for information disclosure shall not be less than 5 working days The competent authority for ecology and environment shall make a decision on the EIA report, according to public opinions and comments, related legislation, standard and norms
Once the decision on the EIA report is made by the competent authority for ecology and environment	Once the competent authority for ecology and environment has made a decision on the EIA report, the decision shall be posted on the website with the information on how to bring an administrative review and administrative appeal

MEP (became the MEE in March 2018) to provincial EPBs; and further empowered to municipal EPBs by provincial EPBs, based on practical local conditions. At present, the EIA documents for 3 types of construction project shall be reviewed and approved by the MEE, including:

① projects of a special nature, such as nuclear facilities or highly confidential projects;
② projects that involve more than one province, autonomous region and municipality directly under the central government (cross-boundary projects);
③ projects approved by the State Council or by related departments authorized by the State Council.

Before 2016, all types of EIA documents, including the EIA report, the EIA Form and the EIRF, shall be reviewed and approved by the competent authority for ecology and environment at various levels. In the EIA Law (2016), the review and approval for Project EIA was further simplified. The review and approval for the EIRF is no longer required but replaced by reporting and archiving. The competent authority for ecology and environment will review the EIA document (the EIA report or EIA form) in consultation with relevant authorities involved in the proposed project. Once the overall conclusions of environmental impact assessment and the measures of ecological and environmental protection are accepted, the EIA document will be approved with a statement to elaborate the likely environmental effects caused by the proposed project and the countermeasures the developer must adopt, including conditions, mitigation measures, the follow-up program, and so on. Should there be any negative comments on the process, the content and the conclusions of the EIA process, or the quality of the EIA document is unacceptable, the EIA document will be rejected with a statement to elucidate the critical issues. However, a revised EIA document resubmitted for further review and approval is permitted.

4.5.6 *EIA Follow-up for Construction Project*

In the process of project construction and operation, if there is any circumstance that is inconsistent with the approved EIA document, the developer shall organize a post-evaluation on environmental impacts, and report to the competent authority for ecology and environment and competent authority for project approval for records and archiving. The competent authority for ecology and environment may also request the developer to conduct the post-evaluation on environmental impacts and to take necessary measures for improvement. In practice, EIA follow-up was mainly implemented in the following industrial sectors:

① Projects of hydrological engineering, hydropower, mining, ports and railway industries where the levels and extents of environmental impacts are much greater and larger, as these major environmental impacts will only gradually emerge after these projects are completed and operated for a certain period of time, as well as projects from other industries, which will generate impacts on important ecological and environmentally sensitive zones.

② Projects of metallurgical, petrochemical and chemical engineering industries with significant environmental risks, sensitive construction sites, and continuous emissions of heavy metals or persistent organic compounds.

There are seven major items within the content of EIA follow-up assessment, as follows:

① Review of the process of the project, including the EIA, the completion and acceptance of environmental protection facilities, the implementation of mitigation measures, environmental monitoring, public participation, and so on;
② Assessment of the engineering process of project, including project site, scale, manufacturing techniques, operation and dispatching modes, the sources of environmental pollutions or ecological impacts, and the patterns, level and extent of the impacts;
③ Assessment of the changes of the regional environment, including the changes of environmentally sensitive targets in the project site surroundings, the changes of pollution sources or other sources of impacts, the status quo of the environmental quality, and the analysis of the variation tendency;
④ Assessment of the effectiveness of mitigation measures, including the applicability and effectiveness of measures of pollution control, ecological protection and risk prevention (as specified in the original EIA report), and the compliance with the requirements of national or local laws, regulations, and standards;
⑤ Verification of the predicted environmental impacts, including the differences between the predicted and actual impacts on key environmental components, major omissions or mistakes within the content and conclusions of the original EIA report, if any, and the appearance of persistent, cumulative and uncertain environmental impacts;
⑥ Remediation and improvement measures of environmental protection; and
⑦ Conclusions of the EIA follow-up (Chang, Wang, Wu, Sun, & Hu, 2018).

4.6 Regional EIA (REIA)

Regional Environmental Impacts Assessment (REIA) is the EIA for regional development activities, such as watershed development, construction of economic and technological development zones, construction of new urban areas, and urban renewal. Before SEA concepts were officially adopted in China, Chinese scholars and practitioners proposed REIAs to evaluate the environmental impacts of regional development plans and practices (Zhu & Ru, 2008). In mid 1980s, theoretical studies on REIA's concepts, meaning, types, contents, procedures, methodologies, and groundwork of assessment were initiated in China. As regulated in Article 4 of *the Interim Regulations on Environmental Management for Economic Zones Open to the Outside World*, promulgated in March 1986, EIA should be conducted for any new district construction in these special economic zones, according to the principles of comprehensive planning and rational layout. This is the first legal provision on REIA.

Since 1989, several areas were selected for REIA pilot study, including Baiyin City (Gansu Province), Meizhou Bay (Fujian Province), Kaiyuan City (Yunnan Province), by the Environmental Quality Assessment Professional Committee of the Chinese Society for Environmental Sciences. In addition, two REIA pilot projects, REIA for Maanshan City (Anhui Province) and REIA for Xigu District of Lanzhou City (Gansu Province) were initiated by the NEPA. In January 1993, basic principles and management procedures for REIA were proposed in *Several Opinions Concerning to further Improve the Environmental Protection and Management for Construction Project*, promulgated by the NEPA. Later on, REIA shall be conducted for watershed development, construction of economic and technological development zones, construction of new urban areas, and urban renewal, as depicted in *the Ordinances of Management for Environmental Protection of Construction Project* promulgated by the State Council in November 1998.

In China, the same technical approach and management approach of Project EIA should be applied to REIA. The main content of REIA including:

① assessing current environmental status;
② predicting the possible environmental impacts of development activities;
③ proposing environmental management plans to minimize negative environmental impacts.

Nevertheless, after the enforcement of the EIA Law in 2002, REIA is gradually replaced by PEIA (EIA for Plan).

4.7 EIA for Plan (PEIA)

Environmental impacts resulted from governmental policies, plans and programs are normally wider, larger and lasting longer than that from construction projects. In order to fully realize environmental pollution prevention and control and ecological conservation at the decision-making stage, and to implement sustainable development strategies, the requirements of EIA was extended to plan, according to the EIA Law enacted in 2002.

4.7.1 Applicable Scope of PEIA

In the EIA Law, only a limited set of planning activities were regulated but policy, program and other planning activities were excluded (Zhu & Ru, 2008). In particular, as specified in the EIA Law, the relevant departments of the State Council, and the local people's governments at (above) the level of the cities with districts as well as their relevant departments shall:

4.7 EIA for Plan (PEIA)

- conduct EIA for, and incorporate a chapter or explanation concerning the environmental impacts into the integrated plans (including the land utilization plans and the construction, development and utilization plans for regions, watersheds, and sea areas), while compiling the plan (Article 7).
- conduct EIA and prepare an EIA report for the special plans (including the development plans of industry, agriculture, animal husbandry, forestry, energy, water conservancy, transportation, urban construction, tourism, and natural resources utilizations), prior to reporting the draft special plan for review and approval (Article 8).

Currently in China, there are no clear definitions of integrated plans and specific plans but characterized through enumeration methods in the EIA Law. The integrated plans deal with issues related to land utilizations and cross boarder matters. The special plans deal with issues subject to a specific sector, but usually with is a list of detailed construction projects within the special plan. To further clarify the applicability of the EIA report and the EIA chapter/explanation, according to Article 9 of the EIA Law, two guidelines, *Scope of the Plans to Prepare the Environmental Impact Report (on Trial)* and *Scope of the Plans to Prepare the Environmental Impact Chapters (on Trial)*, were issued by the SEPA in July 2004.

4.7.2 Roles and Responsibilities

In PEIA, there are 3 major participants, including the planning agency, the review and approval agency, and the appraisal group. Their roles and responsibilities are depicted as followings.

- The Planning Agency: The agency proposed the draft plan shall conduct the EIA for the draft plan and organize the activities of public participation. The EIA task might be undertaken by the planning agency or a commissioned consultancy.
- The Review and Approval Agency: This is a governmental agency superior to the planning agency, which is responsible for reviewing and approving the draft plan, and examining the EIA report and comments acquired from the activities of public participation.
- The Appraisal Group: This is a team to examine the EIA report for the proposed plan and comments from the general public toward the proposed plan. The coordinator of the appraisal group should be the competent authority for ecology and environment or related department appointed by the review and approval agency. The members of the appraisal group are randomly selected from an expert database compiled by the competent authority for ecology and environment.

4.7.3 Approaches and Requirements of PEIA

Various EIA requirements shall be applied to integrated plans and specific plans due to the different inherent characteristics. For integrated plans, EIA shall be performed during the planning preparation phase. In addition, EIA shall be included and conducted as part of the plans. There is no need for an individual EIA report. For specific plans, EIA shall be performed prior to submitting the draft plan for review and approval. In this case, EIA reports shall be submitted for review as a separate task. Various EIA requirements for integrated plans and specific plans are depicted in Table 4.4.

In 2014, *the Technical Guidelines for Environmental Impact Assessment for Plans (on Trial)* (refer to Table 4.1) was issued by the MEP in guiding agencies and EIA practitioners to conduct the PEIA work. The PEIA report should be concise with appropriate illustrations, detailed data, clear arguments and definite conclusions, and so on. The contents of the PEIA report for a special plan and the PEIA chapter for an integrated plan are listed in Box 4.4.

Box 4.4 Contents of the PEIA report for a plan and the EIA chapter for an integrated plan

The PEIA report for a special plan should include the followings:
a. General principles;
b. Overview of the proposed plan;
c. Description of existing environmental conditions;
d. Analysis and assessment on environmental impacts;
e. Recommended programs and mitigation measures;
f. Monitoring and follow-up assessment;
g. Public participation;
h. Difficulties and the uncertainties;
i. Executive summary.
   The PEIA chapter for an integrated plan should include the followings:

a. Introduction;
b. Analysis of existing environmental conditions;
c. Analysis and assessment on environmental impacts;
d. Mitigation measures for environmental impacts.

4.7.4 Public Participation in PEIA

According to Articles 11 of the EIA Law, for specific plans with potential adverse environment impacts and directly involving public environmental rights and interests,

4.7 EIA for Plan (PEIA)

Table 4.4 EIA requirements for integrated plans and specific plans

Item	Integrated plans	Specific plans
Planning agency	The relevant departments of the State Council, and the local people's governments at (above) the level of the cities with districts as well as their relevant departments	
Type of plan	The land utilization plans and the construction, development and utilization plans for regions, watersheds, and sea areas	The development plans of industry, agriculture, animal husbandry, forestry, energy, water conservancy, transportation, urban construction, tourism, and natural resources utilizations
EIA requirements	Prepare and include a chapter or statement on environmental impacts as part of the overall plan. There is no need to prepare a separate EIA report	Prepare a separate EIA report
Content of EIA	Analysis, forecasting and evaluation of possible negative environmental impacts resulting after plan implementation Provide countermeasures to prevent or mitigate negative environmental impacts	Analysis, forecasting and evaluation of possible negative environmental impacts resulting after plan implementation Countermeasures and steps to prevent or mitigate negative environmental impacts Conclusion of EIA
Timing	During the plan preparation process (For integrated/guidance plans, the SEA and the plan should be conducted simultaneously)	After the plan draft is prepared and before it is submitted for review and approval
Public participation	Not required	The planning agency shall held public meetings, hearings, or other activities to solicit comments on draft EIA report from stakeholders prior to submitting draft plans for review and approval
Review of EIS	Not required	Written comments should be provided by a designate review team

the planning agency should organize public meetings, hearings or take other activities to solicit opinions on the draft PEIA report from relevant departments, experts and the public before submitting. The planning agency should then seriously consider opinions on the draft PEIA report and include an explanation on whether opinions on the PEIA report have been adopted or not, along with the submission. However, for plans involving classified information of national security or confidential information, public participation is not required.

4.7.5 Submission and Review of the PEIA Documents

(1) Submission of the PEIA documents

For integrated plans such as including the land utilization plans and the construction, development and utilization plans for regions, watersheds, and sea areas, PEIA should be conducted concurrently while compiling the plans. As a part of the draft plan, the PEIA chapter should be submitted to the review and approval agency. For special plans, PEIA should be conducted prior to submitting the draft plan, and the PEIA report should be submitted independently to the review and approval agency.

(2) Review of the PEIA report of the specific plans

It is difficult for the review and approval agency to undertake detailed technical review on the PEIA report of the draft plan. In order to assure the review process is completely and successfully implemented, as regulated by the EIA Law, an appraisal group consisted of experts and representatives from relevant departments should be organized to conduct professional and technical review on the PEIA report of the draft plan. The coordinator of the appraisal group should be the competent authority for ecology and environment or related department appointed by the review and approval agency. The reviewing process of the PEIA report of the draft plan was elucidated in *the Measures of Reviewing the Environmental Impact Assessment Report of Special Plan* issued by the SEPA in October 2003. A written review comments containing conclusions, rationale and recommendations shall be prepared by the appraisal group.

4.7.6 Approval of Plan

The planning agency shall submit the draft plan, the PEIA report and the review comments from the appraisal group to the review and approval agency. The review and approval agency shall take the conclusion of the PEIA report and the review comments as the important base for their decision-making. In case the conclusion of the PEIA report or any of the review comments are not adopted in the final decision on the proposed plan, an explanation shall be made and shall be kept in archived files for further reference.

4.7.7 EIA Follow-up for Plan

The implementation of a plan is a long-term process. And, it is still quite difficult to ensure that there would not be any new or unexpected environmental problems resulting from the implementation of the plan or the changes on social, economic and natural circumstances, even if with thorough and comprehensive PEIA. It is necessary for the planning agency to conduct follow-up assessment to find out any new or unexpected environmental issues so as to take necessary corresponding countermeasures. As regulated in Article 15 of the EIA Law, after the specific plans with significant environmental impacts have been implemented, the planning agency should organize timely follow-up assessment and submit the evaluation results to the review and approval agency. Once evident adverse environmental impacts are identified, mitigation measures and countermeasures should be implemented immediately.

4.8 Certificate System of EIA Consultancy (Revoked in 2018)

EIA is a professional work involving wide range of techniques and should be scientific, objective, and impartial. The certificate system for EIA consultancies was introduced in 1986 by the NEPA to ensure the quality of EIA work. Through years of practice, the certificate system has been proved to be as an effective means to regulate the implementation of EIA and to promote the quality of EIA. The certificate system of EIA consultancy was concretely founded through the promulgations of *the Ordinances of Management for Environmental Protection of Construction Project (1986, 1998), the Measures of Management for the Certificate System of Environmental Impact Assessment of Construction Project (on Trial) (1986), the Measures of Management for the Qualification and Certificate System of Environmental Impact Assessment of Construction Project (1999, 2005)* and the EIA Law (2002), where the grades, scope, applications, review and approval, auditing and supervision, penalties, and the qualification criteria of EIA certificate were clearly regulated. The provisions regarding the certificate system for EIA consultancy can be summarized in Box 4.5. Along with booming market of EIA consulting business, the number of certified EIA consultancies has been constantly increasing. However, the number of grade A certificates is limited to less than 200, as shown in Fig. 4.3.

Box 4.5 Some provisions concerning the certificate system for EIA consultancy
- *Project EIA task should only be performed by certified consultancy with appropriate certificate issued by the competent authority for ecology and environment of the State Council. No individual shall undertake EIA consultation task.*
- *Qualification criteria and management measures of certified consultancy are regulated by the competent authority for ecology and environment of the State Council.*
- *Certificates are classified into grade A and grade B with defined working scope. Certified consultancy should conduct EIA task in accordance with the grade and scope limit. Certified consultancies with grade A are authorized to*

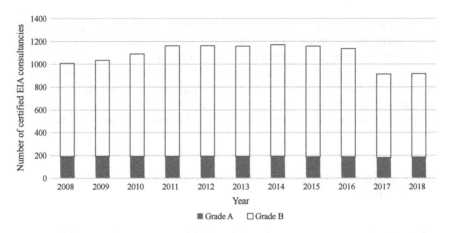

Fig. 4.3 Number of certified EIA consultancies (2008–2018)

> undertake EIA for projects approved by the competent authority for ecology and environment at all levels, while certified consultancies with grade B are authorized to undertake EIA for projects approved by the competent authority for ecology and environment at provincial level or below.
> - *The List of certified consultancies is promulgated by the competent authority for ecology and environment of the State Council. The developer and the public can acquire information about the certified consultancies from the list published on the website.*
> - *In order to assure the independence and impartiality of EIA task, and fair review of the EIA documents, there should be no interest correlations between certified consultancies and the competent authority for ecology and environment or other relevant departments.*

4.9 Management of Qualified EIA Practitioners

In order to establish the qualification system of EIA practitioners, the first session of "Project EIA Training Course for EIA Professionals" was held by the SEPA in Zhejiang Province, in June 1990, and "Certificate of Project EIA Practice Training" was issued to qualified trainees. In 1998, the certificate system of EIA professionals was implemented to require all EIA practitioners to be certified to perform EIA work, as regulated in *the Measures of Management for the Qualification and Certificate System of Environmental Impact Assessment of Construction Project* (1999).

Further, in February 2004, *the Interim Regulations on the System of EIA Engineers Vocational Certificate, the Implementation Measures of the Examination of*

EIA Engineer Vocational Certificate and *the Measures of the Review of EIA Engineer Vocational Certificate* were promulgated by the Ministry of Personnel and the SEPA, jointly, to strengthen the management of EIA professionals, to improve the qualification and operation of EIA practitioners, to guarantee the quality of EIA, and to safeguard national environmental security and public interests. On April 1, 2004, the System of EIA Engineer Vocational Certificate was officially launched to issue vocational certificate to qualified EIA practitioners to undertake Project EIA and technical review of the EIA documents. The examination of EIA engineer vocational certificate is composed of 4 subjects, such as 'Laws and Regulations of EIA', 'Technical Guidelines and Standards of EIA', 'Techniques and Methods of EIA' and 'Case Study on EIA', which is held at the second quarter of the year.

In addition to the System of EIA Engineer Vocational Certificate, the System of Periodic Registration of Vocational Certificate is implemented in which qualified EIA engineer is required to register the vocational certificate at the registration management office within 3 years after obtaining the vocational certificate to become a 'Registered EIA Engineer'. Fail to do so within the specific time frame will result in the automatic revocation of the vocational certificate. Registered EIA engineer should be employed by certified Project EIA agency to be eligible to perform EIA tasks.

4.10 Prospects of EIA in China

In order to improve the effectiveness of EIA and to accumulate experiences and lessons from EIA practices, the Chinese EIA system is continuously evolving and advancing through fine-tuning, adjustment and improvement in the following ways.

(1) To expand the target and scope of SEA, and to integrate the national economic and social development plans and policies into assessment scope

Many environmental problems are caused by the intentional or unintentional negligence of environmental factors during policy making and plan compiling. Among diverse integrated plans, the national economic and social development plans are the most important ones to generate great influences on environment. Therefore, SEA should be applied to the national economic and social development plans to fulfill the win-win situation of economic development and environmental protection, as well as to build a resource conservation and environmental friendly society.

(2) To enhance public participation in EIA, especially for NGOs

It's necessary to encourage and convene more social organizations and the general public to participate in the management of EIA, to exercise the "Right To Know", the "Right To Participate", and the "Right To Supervise", as ensured and guaranteed by laws, and to play active and aggressive roles in all public participation activities, such as demonstration meetings, public hearings, seminars and other forms.

(3) To build a platform of EIA information disclosure through integration of data and information from all relevant institutions

Diverse EIA information or related, including the information regarding the status of the EIA documents, the final decisions on the EIA documents, completion and acceptance of construction projects, and so on, were publicized on the websites of the competent authorities for ecology and environment at various levels. In order to promote public participation in EIA, the disclosed information should be wider and in-depth, such as reviewing comments, reviewing time, project sites, construction methods, key indices in EIA, mitigating measures, and so on. In addition, several coordinating agencies should be established to build and strengthen the cooperation mechanisms among departments, to share big data from all types of businesses and departments, and to set up the sound basic database.

References

Chang, I. Shin, Wang, W., Wu, J., Sun, Y., & Hu, R. (2018). Environmental impact assessment follow-up for projects in China: Institution and practice. *Environmental Impact Assessment Review, 73*, 7–19. https://doi.org/10.1016/j.eiar.2018.06.005. [2019-05-18].

Craik, N. (2008). *The international law of environmental impact assessment. Process, substance and integration.* New York: Cambridge University Press.

Glasson, J., Therivel, R., & Chadwick, A. (2011). *Introduction to environmental impact assessment* (4th ed.). New York: Routledge.

Joseph, C., Gunton, T., & Rutherford, M. (2015). Good practices for environmental assessment. *Impact Assessment and Project Appraisal, 33*(4), 238–254. https://doi.org/10.1080/14615517.2015.1063811. [2019-05-18].

Li, Terry H. Y., Thomas Ng, S., & Skitmore, M. (2012). Public participation in infrastructure and construction projects in China: From an EIA-based to a whole-cycle process. *Habitat International, 36*(1), 47–56. https://doi.org/10.1016/j.habitatint.2011.05.006. [2019-05-18].

Senecal, P., Goldsmith, B., Conover, S., Sadler, B., & Brown, K. (1999). Principles of environmental impact assessment [computer file]. Retrieved from https://www.iaia.org/uploads/pdf/principlesEA_1.pdf. [2019-05-19].

Wang, Y., Morgan, R. K., & Cashmore, M. (2003). Environmental impact assessment of projects in the People's Republic of China: new law, old problems. *Environmental Impact Assessment Review, 23*(5), 543–579. https://doi.org/10.1016/S0195-9255(03)00071-4. [2019-05-19].

Wu, J., Chang, I. Shin, Yilihamu, Q., & Zhou, Yu. (2017). Study on the practice of public participation in environmental impact assessment by environmental non-governmental organizations in China. *Renewable and Sustainable Energy Reviews, 74,* 186–200. https://doi.org/10.1016/j.rser.2017.01.178. [2019-05-19].

Zhang, Y., Liu, X., Yu, Y., Bian, G., Li, Y., & Long, Y. (2012). Challenge of public participation in China's EIA practice. In *Paper presented at the 32nd Annual Meeting of the International Association for Impact Assessment,* Porto, Portugal.

Zhao, Y. (2010). Public participation in China's EIA regime: Rhetoric or reality? *Journal of Environmental Law, 22*(1), 89–123. https://doi.org/10.1093/jel/eqp034. [2019-05-20].

Zhu, D., & Ru, J. (2008). Strategic environmental assessment in China: Motivations, politics, and effectiveness. *Journal of Environmental Management, 88*(4), 615–626. https://doi.org/10.1016/j.jenvman.2007.03.040. [2019-05-20].

Chapter 5
Three Synchronizations System

Introduced by the EP Law (1989) as a supplement to the Environmental Impact Assessment System, the Three Synchronizations System requires that all measures and facilities of pollution prevention and control for construction project should be designed, constructed, and operated synchronously with the design, construction, and operation of the main body of the construction project. The Three Synchronizations System and the Environmental Impact Assessment System are two major instruments for the environmental management of construction projects in China.

5.1 The Development of the Three Synchronizations System

In the spring of 1972, a water contamination incident occurred at the Guanting Reservoir (Beijing), one of two important water sources for Beijing. The incident drew the close attention of the CPC, the Chinese central government and the general public (Luo et al., 2007). According to the *Investigation Report on the Situation of Water Pollution at the Guanting Reservoir*, submitted by the investigation group led by the Beijing "Three-Waste" Management Office (waste gas, waste water, and solid waste), a leading group devoted to reservoir protection, designated by the State Council, was constituted to carry out water pollution abatement and environmental governance at the Guanting Reservoir (Chen, 2012). This was the first action taken by Chinese government to control pollution. In June 1972, as explicitly illustrated in the *Report Concerning the Status Quo of Water Pollution at the Guanting Reservoir and Recommended Solutions*, prepared by the SPC and the MHURD, the State Council, the construction of the factory, and three-waste utilizations and management was to be designed, constructed and operated simultaneously.

As clearly stipulated in *Some Regulations Concerning Environmental Protection and Improvement*, which was approved by the State Council in August 1973, for any new construction, reconstruction, and expansion projects, all pollution prevention and control facilities were to be designed, constructed, and operated simultaneously with the design, construction, and operation of the main body of the project; the competent authority for environmental protection, public health, and related areas was expected to carefully examine the design of the pollution prevention and control facilities, oversee their completion and acceptance, and monitor their performance. Thereafter, the Three Synchronizations became China's first environmental management system.

In 1976, the *Report on Strengthening the Environmental Protection Work* was approved by the CCCPC, in which the Three Synchronizations System was reiterated for its necessity and coerciveness. As further clarified in this report, any project that failed to comply with the Three Synchronizations System would not be allowed to be constructed or put into production. Further, the Three Synchronizations System was legally confirmed in the *Law of the People's Republic of China on Environmental Protection (on Trial)* promulgated in September 1979. As stipulated in Article 6, in planning any new construction, reconstruction, and expansion project, the developer would have to submit an EIA report to the competent authority for environmental protection and other relevant agencies for review and approval, prior to the designing of the project. The pollution prevention and control facilities would have to be designed, constructed, and operated simultaneously with the main body of the project. The discharge of any kind of harmful substances would have to be in compliance with state regulations and national standards.

In 1977, a special investigation on the implementation of the Three Synchronizations in 23 provinces and cities was conducted by the State Council. It revealed totally unsatisfactory result at all levels owing to the fact that only the principles of the Three Synchronizations were specified in relevant laws and regulations without any definite and detailed legal provisions regarding management systems, institutional responsibilities, authority, review and approval procedures, and legal liabilities in particular. In order to counteract this reality, the *Decisions on Strengthening the Environmental Protection Work During the Period of National Economy Adjustment* was issued by the State Council in February 1981 to reemphasize the importance of the Three Synchronizations System and to enact relatively clear requirements regarding the investment in and equipment required for pollution prevention and control facilities. Additionally, in May 1981, the *Measures of Management for Environmental Protection of Infrastructure Construction Project* was jointly promulgated by the SPC, SCC, SEC and LGEP to further institute more comprehensive and specific regulations on the requirements, management procedures, and penalties of the Three Synchronization System. Therefore, the Three Synchronizations System was made a reality and included in construction procedures.

In March 1986, the *Measures of Management for Environmental Protection of Construction Project* was jointly publicized by the EPC, SPC, and the SEC as a replacement for the *Measures of Management for Environmental Protection of Infrastructure Construction Project*, based upon the practical experiences and lessons learned from the implementation, in which the specific content of the Three Synchronizations System was established. In March 1987, the *Regulations on the Design of Environmental Protection of Construction Project* was jointly promulgated by the SPC and EPC to further supplement and improve the Three Synchronizations System (Sinkule, 1994).

5.1 The Development of the Three Synchronizations System

Based upon the experience accumulated from the implementation of the Three Synchronizations System, the *Law of the People's Republic of China on Environmental Protection* was promulgated in December 1989 to further confirm the legal status of the Three Synchronization System, as stipulated in Article 26: "The pollution prevention and control facilities in construction project must be designed, constructed and operated simultaneously with the main body of the project. Once have past the completion and acceptance supervised by the competence authority for environmental protection, the pollution prevention and control facilities shall be put into operation." In addition, in overcoming the problem of poor enforcement of the Three Synchronization System, the following regulation was included in Article 26: "The pollution prevention and control facilities cannot be dismantled or idled. If it is necessary to dismantle or idle the facilities, the approval from the competence authority for environmental protection must be obtained." Besides, the legal liability for noncompliance with the Three Synchronizations System was clearly regulated in Article 36. The Three Synchronizations System was further strengthened and improved through the promulgation of the *Ordinances of Management for Environmental Protection of Construction Project* in November 1998 and the *Measures of Management for the Completion and Acceptance of Environmental Protection of Construction Projects* in December 2001.

The Three Synchronizations System has played an important role in stimulating investment in the pollution prevention and control facilities of construction projects. However, problems still exist in many instances as the procedures of the Three Synchronizations System were not strictly abided by. Should the construction project operate without the approval from the local EPB, the owner of the construction project shall be punished. In 2004, about 62.35% of construction projects (79,500 out of 127,500) were subjected to compliance with the Three Synchronizations procedures, whereby the Three Synchronizations procedure of nearly 95.60% of these construction projects (over 76,000 out of 79,500) were approved. However, there was evidence that the penalties resulting from noncompliance with the Three Synchronizations procedure were not strictly enforced by a few of local authorities. To surmount this problem, a deposit-refund mechanism for the Three Synchronizations System of construction projects has been introduced in some areas. A deposit, calculated on the basis of the total investment cost of the construction project, can be returned to the developer upon fulfillment of the requirements of the Three Synchronizations System. However, there is no adequate legal basis for the deposit-refund mechanism or clear criteria for evaluation and returning deposits (Li & Michalak, 2012).

Since the 18th NCCPC held in November 2012, with the transformation of government functions and the deepening of reforms to the administrative review and approval system, the combination of streamlining administration and delegating power to lower levels has become an inevitable trend of national administration. The procedure of the Three Synchronizations System was simplified by revoking the requirements of administrative review and approval from the competent authority for environmental protection. However, the developer shall bear the responsibility for inspecting the completion and acceptance of the environmental protection facilities of the construction project, preparing the inspection report on the completion and acceptance of the environmental protection facilities, and posting related information on the website. These distinct changes can be located in the amendments of several laws and regulations, such as the *EP Law* (amended in 2014), the *EIA Law* (amended in 2016, 2018), and the *Measures of Management for Environmental Protection of Construction Project* (amended in 2017).

5.2 The Requirements of the Three Synchronizations System

The timing and functions of several major environmental management systems for construction projects are shown in Fig. 5.1. The Three Synchronizations System is implemented throughout the lifecycle of the projects, but it is limited to the design, construction and operation of environmental protection facilities of the projects (Chang, Wang, Wu, Sun, & Hu, 2018).

5.2.1 Synchronous Design

In the proposal stage, the possible impacts resulting from the proposed construction project was to be briefly described. In the feasibility study (design specification) stage, specific discussions on environmental protection should be integrated in the feasibility study report to include environmental conditions around the proposed construction project, main pollution sources, major pollutants, possible ecological

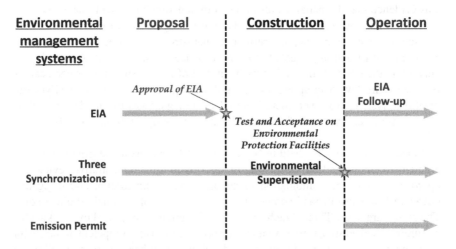

Fig. 5.1 Different timing for various environmental management systems

changes caused by development, preliminary schemes of pollution control, estimation of environmental protection investment, applicable environmental standards to be adopted, and so on. A chapter on environmental protection was to be included in the preliminary design to demonstrate the basis of environmental protection design, main pollution sources, major pollutants and the means of discharge, the applicable environmental standards to be adopted, a brief description of environmental protection facilities to be installed along with their engineering processes, measures for preventing ecological changes, the estimation of environmental protection investment, and so forth.

5.2.2 *Synchronous Construction*

In the construction stage, environmental protection facilities must be constructed along with the main body of the project. During the construction process, it is necessary to protect the surrounding environment at the construction site, to prevent unexpected damage to the natural environment, and to prevent and minimize the pollution and harm to the surrounding environment from dust, noise, vibration, and so on. The competent authority for environmental protection shall carry out the on-site inspection to ensure that the construction process is in compliance with related environmental legislations.

5.2.3 *Synchronous Operation*

Prior to formal production or operation of the construction project, the developer is expected to submit the "Completion and Acceptance Report of Environmental Protection Facilities", with a description of the operation conditions, the results of environmental monitoring and the applicable environmental standards, to the competent authority for environmental protection for review and approval. Since the promulgation of the Interim Measures of the Completion and Acceptance of Environmental Protection of Construction Projects in November 2017, there is no need to seek the review and approval of the "Completion and Acceptance Report of Environmental Protection Facilities" from the competent authority for environmental protection. The developer is expected to post the report on the website to accept the supervision from the general public and the authorities.

5.2.4 *Roles and Responsibilities*

The competent authority for environmental protection is expected to monitor and supervise the environmental protection of construction projects, including auditing the relevant content of environmental protection in the design specifications (feasibility study report); reviewing and approving the EIA documents; inspecting the environmental protection activities during the construction process; inspecting and supervising the operation of the environmental protection facilities; and determining any violations of the Three Synchronizations procedures and enforcing any ensuing penalty. The developer is expected to prepare the EIA report, carry out the environmental protection measures proposed in the preliminary design, and ensure the normal operation of the environmental protection facilities after the completion of the construction project.

5.2.5 The Violation

For any violations of the provisions of the Three Synchronizations, the developer shall be punished and fined. The competent authority for environmental protection shall record any information related to environmental violations of construction projects in the social integrity archives and post the list of offenders on the website.

References

Chang, I. Shin, Wang, W., Wu, J., Sun, Y., & Hu, R. (2018). Environmental impact assessment follow-up for projects in China: Institution and practice. *Environmental Impact Assessment Review, 73,* 7–19. https://doi.org/10.1016/j.eiar.2018.06.005. [2019-06-08].

Chen, Q. (2012). The sustainable economic growth, urbanization and environmental protection in China. *Forum on Public Policy: A Journal of the Oxford Round Table.*

Li, W., & Michalak, K. (2012). Environmental compliance and enforcement in China. In A. Tarantino (Ed.), *Governance, risk, and compliance handbook* (pp. 379–391). Wiley.

Luo, W., Wang, T., Lu, Y., Giesy, J. P., Shi, Y., Zheng, Y., et al. (2007). Landscape ecology of the Guanting reservoir, Beijing, China: Multivariate and geostatistical analyses of metals in soils. *Environmental Pollution, 146*(2), 567–576. https://doi.org/10.1016/j.envpol.2006.08.001. [2019-06-08].

Sinkule, B. J. (1994). Implementation of China's three synchronizations policy: Case studies of wastewater treatment measures at new and renovated factories.

Chapter 6
Emissions Charges System and Environmental Protection Tax System

On December 25, 2016, *the Law of the People's Republic of China on Environmental Protection Tax* (the EPT Law) was promulgated (and came into force on January 1, 2018), with the intention of serving as a replacement for the Emissions Charges System, which had been implemented for over 30 years. The enactment of the EPT Law is not only a milestone for the implementation of the statutory taxation principle, but also an important manifestation of China's determination to carry out an economic transformation leading toward greener development.

6.1 Development of the Emissions Charges System

In December 1978, the Emissions Charges System in China was first proposed in the *Key Points of Environmental Protection Work Report*, submitted by the LGEP and endorsed by the CCCPC, by stipulating that the competent authority for environmental protection shall, in conjunction with other departments concerned, formulate specific measures of the Emissions Charges System. The following was explicitly stipulated in the EP Law (on Trial) promulgated in August 1979: "Discharge of waste gas, waste water and waste residues shall be in compliance with the standards set by the State. When the emission of pollutants exceeds the limits of the standards, emissions charges shall be paid according to the quantity and concentration of the pollutants released." Subsequently, some pilot projects were launched in several provinces to initiate the pilot stage of the Emissions Charges System.

In February 1982, *the Interim Measures of Emissions Charges Collection* was enacted by the State Council, based on the practical experience of the above-mentioned pilot programs. As explicitly stipulated in Article 6, 'for any new construction, reconstruction, and expansion projects with pollutant discharges that exceed the emission standards, …their emissions charges should be doubled; … and the people's government at provincial level are empowered to make appropriate adjustment regarding the amount of emissions charges according the practical conditions.' Further, it was illustrated in Article 10 that concrete implementation measures shall be instituted by the people's governments at the provincial level, according to this Interim Measures. The Emissions Charges System was formally established as the *Interim Measures of Emissions Charges Collection* came into force on July 01, 1982, and was widely implemented throughout the nation. The Emissions Charges System was also specifically stipulated in the *Law of the People's Republic of China on Marine Environment Protection* (1982), the *Law of the People's Republic of China on Water Pollution Prevention and Control* (1984), and the *Law of the People's Republic of China on Air Pollution Prevention and Control* (1987).

The Emissions Charges System was established as a supplementary command and control tool for environmental management, as well as a mechanism for raising capital to compensate for the insufficient budget for environmental administration and enforcement programs (Wu & Tal, 2018). Accordingly, the *Measures of the Financial Management and Accounting for the Emissions Charges* was jointly issued in May 1984 by the MOF and the Ministry of Urban-Rural Development and Environmental Protection (reorganized as the Ministry of Construction in May 1988, and elevated as the MHURD), to regulate the budget management, budge terms, payment settlement, and accounting measures of emissions charges. The emissions charges collected were integrated into the environmental protection subsidy funds to be used for key pollution sources abatement, regional pollution prevention and control, and capacity building for local institutions and authorities for environmental protection.

In July 1988, the *Interim Measures of the Compensable Utilizations of the Special Funds for Pollution Source Governance* was issued by the State Council to significantly change the utilizations of the special funds, from appropriations to loans. Later on, with the continuous economic development and emergence of new environmental problems, a series of policies concerning the utilization and management for the emissions charges were proposed and implemented.

In order to control the increasingly serious damage from acid rain, in September 1992, the *Notice on the Implementation of the Pilot Project to Collect Emissions Charges of Sulfur Dioxide from Industrial Coal Burning* was jointly promulgated by the SEPA, State Price Bureau, MOF, and Economy and Trade Office of the State Council to specifically and importantly expand the scope of emissions charges. Prior to 1992, only those exceeding the emission standards were required to pay emissions charges. However, the promulgation of this regulation initiated the process of collecting emission charges from targeted pollution activities that complied with emission standards. In 1993, *the Notice on Collecting the Emissions Charges of Sewage* was jointly issued by the SPC and the MOF to start the collection of emissions charges from sewage discharge activities that are in compliance with emission standards. Several decrees and regulations related to the emissions charges of sulfur dioxide were promulgated thereafter, as listed in 3 6.1.

> **Box 6.1 Some decrees and regulations related to the emissions charges of sulfur dioxide**
> - *The Rescriptum on the Issues Concerning the Expansion of the Pilot Project to Collect Emissions Charges of Sulfur Dioxide (the State Council [1996] No. 24).*
> - *The Notice on the Expansion of the Pilot Project to Collect Emissions Charges of Sulfur Dioxide in Acid Rain Control Zones and Sulfur Dioxide Pollution Control Zones (Two-Control-Zone) (the NEPA, the SPC, the MOF, and the State Economy and Trade Commission [1998] No. 6).*
> - *The Notice on the Implementation of Charge rates in the Pilot Project to Collect Emissions Charges of Sulfur Dioxide (the SEPA [2000] No. 75).*

Based on 20 years of practical experience, the Emissions Charges System was formalized through the promulgation of *the Ordinance on the Collection, Utilizations and Management for the Emissions Charges*, by the State Council in January 2003, explicitly declaring that the compliance charges and noncompliance charges shall be both executed as a replacement for the initial noncompliance charges, for various pollutants; the emissions charges must be included in the financial budget and incorporated into the special funds of environmental protection for management; and the emissions charges must be used for the pollution prevention and control of key pollution sources, regional pollution prevention and control, as well as the development, demonstration and application of new technologies and techniques of pollution prevention and control. In February 2003, the Management Measures of the Emissions Charges Standard was enacted by the SDPC, MOF, SEPA and SETC to further regulate the standards of emissions charges. In September 2014, the *Notice on the Issues of Adjusting the Emissions Charges Standards* was issued by the NDRC, the MOF, and the MEP to further improve the emissions charges standards, in which emissions charges rates were doubled to significantly increase the economic consequences of polluting activities (Wu & Tal, 2018). Furthermore, it was also specified that the polluter and its manager shall be prosecuted for noncompliance with emission standards.

However, in order to cope with rapid economic development and increasing industrial activities, the EPT Law was enacted on December 25, 2016, and on January 1, 2018 Environmental Protection Tax System effectively replaced the Emissions Charges System.

6.2 The Content of the Emissions Charges System

6.2.1 The Payer

Any firm that directly discharges pollutants (including air pollutants, water pollutants, noise and solid wastes) into the environment, hereinafter referred to as the "polluter", shall pay the emissions charges. However, polluters who have paid the sewage treatment fees for discharging sewage into the centralized municipal wastewater treatment facilities shall no longer need to pay the emissions charges. A polluter who has built facilities and places to deposit, treat, or dispose industrial solid waste in compliance with relevant environmental protection standards does not have to pay the emissions charges.

6.2.2 Category and Amount of Pollutant Emission

The polluter must declare the category and amount of pollutant(s) discharged and provide relevant information to the local competent authority for environmental protection that shall ratify the declaration of the category and amount of the pollutant(s) discharged by the polluter, according to the limits of authority stipulated by the SEPA. A written ratification notice shall be issued to the polluter. Furthermore, the information concerning the category and amount of pollutant(s) discharged shall be disclosed to the public via an internet site.

6.2.3 Category of the Emissions Charges

① According to the *Law of the People's Republic of China on Air Pollution Prevention and Control* and *the Law of the People's Republic of China on Marine Environment Protection*, a polluter who discharges pollutant(s) into the atmosphere and ocean shall pay the emissions charges according to the category and amount of the pollutant(s) discharged.
② According to the *Law of the People's Republic of China on Water Pollution Prevention and Control*, a polluter who discharges pollutant(s) into water shall pay the emissions charges according to the category and amount of the pollutant(s) discharged; if the amount of the pollutant(s) discharged into water exceeds the national or local standards, the emissions charges shall be doubled according to the category and amount of the pollutant(s) discharged.
③ According to the *Law of the People's Republic of China on Solid Waste Pollution Prevention and Control*, a polluter who does note stablish any facilities and places for the storage, treatment, or disposal of industrial solid waste, or if these facilities and places for the storage, treatment, or disposal of industrial solid waste

cannot meet the requirements of relevant environmental protection standards, then the polluter shall pay the emissions charges according to the categories and amount of the pollutant(s) discharged; a polluter who adopts landfill for hazardous waste disposal and cannot meet the requirements of relevant national regulations shall pay the emissions charges according to the category and amount of the pollutant(s) discharged.

④ According to the *Law of the People's Republic of China on Environmental Noise Pollution Prevention and Control*, a polluter who is responsible for environmental noise pollution that exceeds the national noise standards shall pay the emissions charges according to the extent of their exceeding the national noise standards; however, a polluter with paid emissions charges will not be exempted from the liability of pollution prevention and control, compensation for pollution damages, and other responsibilities stipulated by law and administrative regulations.

6.2.4 Charge Rates

Prior to 2003, as each province generally carried out the emissions charges standard recommended by the State, there was only one emissions charges standard. However, the emissions charges standards were gradually changed through the promulgation of various regulations and emission standards. For example, the SO_2 emissions charges standard showed a continuous increase, as shown in Table 6.1. However, due to low emissions charges standards, the total emissions charges need to be raised to enhance the effectiveness of an economic incentive for emission reduction (Bohm, Ge, Russell, Wang, & Yang, 1998; Wu & Tal, 2018).

According to *the Ordinances of Management for the Collection and Utilizations of Emission Charges (2003)*, the people's governments at the provincial level may institute more stringent emissions charges standards according to local conditions. Ever since, various provincial emissions charges standards were promulgated by various local people's governments. By 2016, diverse emissions charges standards for different pollutants were promulgated, as shown in Table 6.2.

6.2.5 Collection and Utilizations of the Emissions Charges

The amount of emissions charges collected has been gradually increasing since the initiation of the Emissions Charges System in 1979, showing a significant uprise in the early 21st century, as displayed in Fig. 6.1. The collection and utilization of the emissions charges must strictly adhere to the concept of "independent accounting for revenues and expenditures". The emissions charges collected at all levels shall be turned over to the state and local government (provincial level) treasuries at the ratio of 1:9. The indispensable expenditures of law enforcement for environmental

Table 6.1 Changes of emissions charges standard for SO_2

Period	Emission standards	Regulations on emissions charges	Charge rate/(CNY/kg)
1982–1992[a]	The Emission Standards of Industrial "Three-Waste" (on Trial) (GBJ 4—1973)	The Interim Measures of Emissions Charges Collection (1982)	0.04
1993–1996[b]		The Notice on the Implementation of the Pilot Project to Collect Emissions Charges of Sulfur Dioxide from Industrial Coal Burning (1992)	0.20
1997–2002[b]	The Comprehensive Emission Standards of Air Pollutants (GB 16297—1996)		
2003–2005[c]		The Ordinance on the Collection, Utilizations and Management for Emissions Charges (2003)	0.20–0.60
2007–2014		The Notice on the Comprehensive Schemes of Work for Energy Conservation and Emission Reduction (2007)	0.60–1.20
2015–2018		The Notice on the Issues of Adjusting the Emissions Charges Standards (2014)	1.20–12.00

[a]During this period, the emissions charges were only collected from pollutants discharged exceeding the emission standards
[b]In the pilot areas, the emissions charges were collected according to the total amount of pollutants discharged, and the pollutants discharged exceeding the emission standards. In other areas, the emissions charges were only collected from pollutants discharged exceeding the emission standards
[c]Since 2003, the emissions charges were only collected according to the total amount of pollutants discharged. It will be illegal for any pollutants discharged exceeding the emission standards and the polluter and its manager shall be prosecuted for violating emission standards

protection shall be included in the departmental budget and guaranteed by the finance budget at the corresponding level. The emissions charges must be included in the financial budget and incorporated into the special funds of environmental protection for management; and the emissions charges must be used for the prevention and control of key pollution sources, regional pollution prevention and control, and the development, demonstration and application of new technologies and techniques of pollution prevention and control, and as the grants and subsidies or soft loans for other pollution prevention and control projects specified by the State Council.

Table 6.2 Emissions charges standards for different pollutants promulgated by various provinces/cities by 2016

District	Pollutant			
	COD	Ammonia nitrogen	SO_2	NO_x
Unit: CNY/kg				
Beijing	10.0	12.0	10.0	10.0
Tianjin	7.5	9.5	6.3	8.5
Shanghai	3.0	3.0	4.0	4.0
Unit: CNY/NP$_{EW}^a$				
Hebei province	2.8	2.8	2.4	2.4
Jiangsu province	4.2	4.2	3.6	3.6
Shandong province	1.4	1.4	3	3
Other provinces	1.4	1.4	1.2	1.2

^{a}NP$_{EW}$: Number of Pollutant Equivalent Weight = Pollutant Discharged/Equivalent Weight of Pollutant (P_{EW})

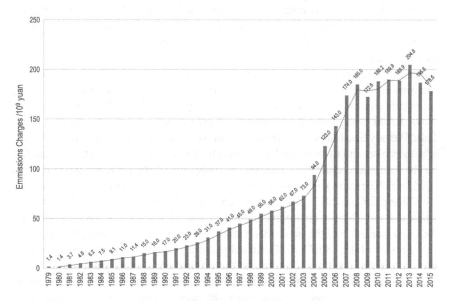

Fig. 6.1 Emissions charges collected from 1979–2015

6.3 The Environmental Protection Tax System

Though continuously improved and enhanced through practices, there are many deficiencies within the Emissions Charges System still to be overcome, as shown in Box 6.2. In addition, there was no legal liability imposed on any parties in non-compliance with relevant laws and regulations to strongly pressurize the polluter(s) to reduce emission or remediate pollution, actively and effectively. Furthermore, the ability of the Central Government to regulate the emissions charges was quite limited as the major portion of the emissions charges was included in the local finance budget, over which the Central Government did not have any jurisdiction.

> **Box 6.2 Major problems within the Emissions Charges System**
> - *The pollutants covered by the System were quite limited, while some important pollutants (such as VOCs) were exempted from the Emissions Charges System.*
> - *Emissions charges were quite cheap in comparison to the costs of pollution prevention and control and pollution abatement to significantly diminish the effectiveness of economic incentive for emission reduction.*
> - *Lacking of capacity of monitoring, supervision and enforcement.*

The idea of introducing the Environmental Protection Tax System to replace the Emissions Charges System was initially suggested in the early 1990s (Bohm et al., 1998). In 2007, the *Notice on the Comprehensive Schemes of Work for Energy Conservation and Emission Reduction* was enacted by the State Council to explicitly propose that it is necessary to enhance and improve the tax policy to actively encourage energy conservation and emissions reduction; thus, the agenda of collecting the environmental protection tax was brought into focus, for the first time. Regarding tax reform and tax management, it was clearly proposed in the *Decision on Several Major Issues Concerning the Comprehensively Deepening Reform*, approved by the Third Plenary Session of the 18th CCCPC in November 2013, to speed up the reform of resource tax and to promote the "environmental protection tax" as the substitution for the "emissions charges". After the promulgation of the EP Law (2014) in April 2014, which became known as the most stringent Chinese environmental protection law (Zhang & Cao, 2015), the enactment of the EPT Law in December 2016 marked a milestone in Chinese environmental legislation. Moreover, it further consolidated the implementation of the statutory taxation principle proposed during the Third Plenary Session of the 18th CCCPC. Actually, the environmental protection tax is not a new category of tax but a substitute for the original emissions charges. In principle, the tax burden of firms would not be increased, as certain preferential tax policies were granted to encourage firms to undertake environment-friendly activities (Jiang & Lu, 2018). The process of evolution from the Emissions Charges System to the Environmental Protection Tax System is displayed in Fig. 6.2, and the comparison between these two systems is shown in Table 6.3.

6.3 The Environmental Protection Tax System

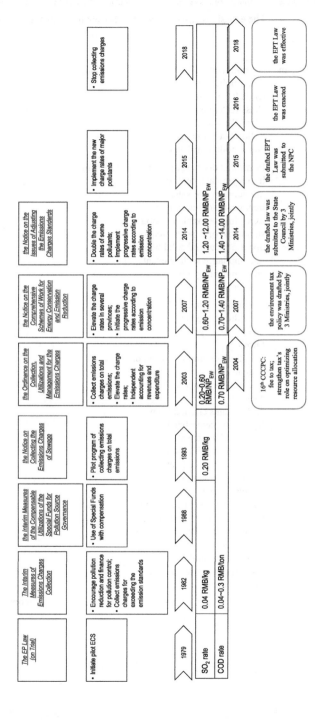

Fig. 6.2 Evolving process from emissions charges to environmental protection tax in China

Table 6.3 Comparison of the emissions charges system and the environmental protection tax system

	The emissions charges system	The environmental protection tax system
Scope		
Pollution categories and item	Air pollutants: 44 items Water pollutants Class I: 10 items Class II: 51 items Class III: 4 items Solid waste: 4 items Noise pollution: 1 items	Unchanged
Exempted firms	Firms that discharge sewage into legitimate, centralized municipal wastewater treatment facilities without exceeding relevant emission standards; or deposit, treat or dispose industrial solid waste in compliance with relevant environmental protection standards	Unchanged
Exempted pollutants	1. Pollutants emitted from agricultural production (excluding large-scale farming) 2. Pollutants emitted by motor vehicles, locomotives, ships, aircrafts and other mobile sources 3. Pollutants emitted from legitimate, centralized municipal wastewater treatment facilities without exceeding relevant national and local standards 4. Solid waste comprehensively utilized by the polluter in compliance with relevant national or local standards 5. Any other pollutants specifically exempted by the State Council	Unchanged

(continued)

6.3 The Environmental Protection Tax System

Table 6.3 (continued)

	The emissions charges system	The environmental protection tax system
Charges		
Charge rates	Flat charge rates for all categories	1. Unchanged flat charge rates for solid waste and noise pollution 2. Local governments can set their own charge rates for air and water pollutant with maximum charge rates of up to ten times the baseline rates (equal to the old flat charge rates)
Chargeable pollutants	Three pollutants with the highest amount discharged per discharge outlet in each category	Local government can increase the number of pollutants per category to charge taxes, and no maximum number is specified
Compliance		
Measurement/reporting	Reported by firms, often without much supervision and verification from local competent authorities for environmental protection and leading to widespread false reporting	Reported by automatic monitoring equipment or networks installed by the competent authorities for environmental protection or the authorized third parties
Governing bodies	The competent authorities for environmental protection	The competent authorities for environmental protection and tax administrations (the tax administrations usually have comparatively greater enforcement power than that of the competent authorities for environmental protection)
Penalties	Noncompliance will result in administrative penalties (mainly fines)	1. Noncompliance will be regarded as tax fraud to result in criminal prosecution 2. Fines are significantly increased from up to three to five times of the original levels

(continued)

Table 6.3 (continued)

	The emissions charges system	The environmental protection tax system
Incentives		
Low-emission incentives	A 50% reduction of emissions charges for firms to discharge applicable air or water pollutants at a concentration level less than 50% of prevailing national or local standards	1. A 50% reduction of emissions charges for firms to discharge applicable air or water pollutants at a concentration level less than 50% of prevailing national or local standards 2. 25% tax reduction if the discharged pollutant concentration is 30%–49% below standards

6.3.1 Who Pays the Environmental Protection Tax

As defined in the EPT Law, whoever, including companies, public institutions and other production operators, discharges chargeable pollutants into the environment, directly, within the territory of China and the sea areas under its jurisdiction, shall pay the environmental protection tax (EPT). However, those, including companies, public institutions and other production operators, who discharge chargeable pollutants into the centralized municipal wastewater treatment facilities and solid waste disposal facilities, or deposit, treat or dispose of solid waste at legally established facilities/sites, are not subject to the EPT, as they do not directly discharge chargeable pollutants into the environment. In addition, a number of pollution sources eligible for the EPT exemption were listed in the EPT Law, including, agricultural sources (excluding large-scale livestock farming); mobile sources (motor vehicles, railroad vehicles, off-road vehicles, mobile machinery, shipping vessels, aircrafts); domestic sewage treatment plants and domestic waste treatment facilities (as long as they are in compliance with national/local discharge rates).

6.3.2 Items Subject to the Environmental Protection Tax

Four categories of pollutants, including air pollutant, water pollutant, solid waste and noise pollution, are subject to the EPT. Carbon dioxide, the earlier hotspot of argument, is not included in the final EPT Law. For any discharge outlet, according to the amount of pollutant discharged (NP_{EW}, number of pollutant equivalent weight), the EPT shall be calculated and collected on the top three air pollutants, top five water pollutants for Class I, and top three water pollutants for other classes. In other words, not all of the pollutants discharged will be charged. A list of chargeable water

6.3 The Environmental Protection Tax System

Table 6.4 Water pollutants (Class I and II) and equivalent weight of pollutant

Category	Water pollutants	Equivalent weight of pollutant, P_{EW}/kg
Class I	Mercury	0.5×10^{-3}
	Cadmium	0.005
	Chromium	0.04
	Hexavalent chromium	0.02
	Arsenic	0.02
	Lead	0.025
	Nickel	0.025
	Benzo (a) pyrene	0.3×10^{-6}
	Beryllium	0.01
	Silver	0.02
Class II	Suspended material (SS)	4
	Biochemical oxygen demand (BOD_5)	0.5
	Chemical oxygen demand (COD_{cr})	1
	Total organic carbon (TOC)	0.49
	Petroleum	0.1
	Animal and vegetable oils	0.16
	Volatile phenol	0.08
	Cyanide	0.05
	Sulfide	0.125
	Ammonia nitrogen	0.8
	Fluoride	0.5
	Formaldehyde	0.125
	Aniline	0.2
	Nitrobenzenes	0.2
	Anionic surfactant (LAS)	0.2
	Copper	0.1
	Zinc	0.2
	Manganese	0.2
	Color developers (CD-2)	0.2
	Phosphorus	0.25
	Elemental phosphorus (P)	0.05
	Organophosphorus pesticides (in P)	0.05
	Dimethoate	0.05
	Methyl parathion	0.05
	Malathion	0.05
	Parathion	0.05

(continued)

Table 6.4 (continued)

Category	Water pollutants	Equivalent weight of pollutant, P_{EW}/kg
	Pentachlorophenol and sodium pentachlorophenol (calculated as pentachlorophenol)	0.25
	Trichloromethane	0.04
	Adsorptive organic halides (AOX) (measured as Cl)	0.25
	Carbon tetrachloride	0.04
	Trichloroethylene	0.04
	Tetrachloroethylene	0.04
	Benzene	0.02
	Toluene	0.02
	Ethylbenzene	0.02
	Ortho-xylene	0.02
	p-xylene	0.02
	m-xylene	0.02
	Chlorobenzene	0.02
	O-dichlorobenzene	0.02
	Dichlorobenzene	0.02
	Nitrochlorobenzene	0.02
	2,4-Dinitrochlorobenzene	0.02
	Phenol	0.02
	m-cresol	0.02
	2,4-Dichlorophenol	0.02
	2,4,6-Trichlorophenol	0.02
	Dibutyl phthalate	0.02
	Dioctyl phthalate	0.02
	Acrylonitrile	0.125
	Selenium	0.02

pollutants and their equivalent weights, within different categories, were shown in Table 6.4.

6.3.3 Basis and Rate of the Environmental Protection Tax

The EPT shall be calculated based on the amount of pollutants discharged. For air and water pollutants, the basis shall be determined according to the number of pollutant equivalent weight. For solid waste, the basis shall be determined according to the quantity discharged. For noise pollution, the basis shall be determined according to

6.3 The Environmental Protection Tax System

Table 6.5 Basis and rate of the environmental protection tax stipulated in the EPT law

Pollutants		Charge rates
Air pollutants		1.2–12 CNY/NP_{EW}
Water pollutants		1.4–14 CNY/NP_{EW}
Solid waste	Coal gangue	5CNY/ton
	Tailings	15 CNY/ton
	Hazardous waste	1000 CNY/ton
	Smelt slag, fly ash, slag, other solid waste (including semi-solid, liquid waste)	25 CNY/ton
Noise pollution		350 CNY/month (exceed 1–3 dB)
		700 CNY/month (exceed 4–6 dB)
		1400 CNY/month (exceed 7–9 dB)
		2800 CNY/month (exceed 10–12 dB)
		5600 CNY/month (exceed 13–15 dB)
		11200 CNY/month (exceed more than 16 dB)

the decibel level exceeding the noise standards, as show in Table 6.5. In addition, the rates of various pollutants were specified in the EPT Law. Especially for air and water pollutants, the rate charged varies from the minimum rate to a ceiling of ten times the minimum rate. Local governments have the full discretion to determine the rates charged locally within the range specified in the EPT Law.

6.3.4 Preferential Policy

If the discharged concentration of the chargeable air pollutants or water pollutants is less than 70% but greater than 50% of the stipulated emission standards, a 25% deduction on the payable amount of the EPT will be granted; if the discharged concentration is less than 50% of stipulated emission standards, then a 50% deduction on the payable amount of the EPT will be granted.

6.3.5 Monitoring and Enforcement

The collection and administration management for the EPT rely heavily on self-reporting by the polluters. Required data are submitted to the competent authorities for tax administrations to determine the appropriate amount of tax. Emission monitoring is supervised by the competent authorities for environmental protection. Detailed emission inventories and data shall be shared between the competent authorities for environmental protection and tax administration. Regarding the legal liabilities, in addition to relevant penalties stipulated in the *Law of the People's Republic of China on Tax Collection and Management* and the EPT Law, the polluter's liability for the damage caused by the discharged pollutants is also stipulated in the EPT Law.

References

Bohm, R. A., Ge, C., Russell, M., Wang, J., & Yang, J. (1998). Environmental Taxes China's Bold Initiative. *Environment: Science and Policy for Sustainable Development, 40*(7), 10–38.

Jiang, S., & Lu, Q. (2018). New measure of environmental protection in China. *The Lancet Planetary Health, 2*(12), e517.

Wu, J., & Tal, A. (2018). From pollution charge to environmental protection tax: a comparative analysis of the potential and limitations of China's new environmental policy initiative. *Journal of Comparative Policy Analysis: Research and Practice, 20*(2), 223–236.

Zhang, B., & Cao, C. (2015). Policy: Four gaps in China's new environmental law. *Nature, 517*(7535), 433–434.

Chapter 7
Target Responsibility System of Environmental Protection and Performance Evaluation System

The Target Responsibility System is a crucial institutional mechanism that enables the central government to monitor, supervise and control the actions of local governments in compliance with the overall planning of the central government (Lo, 2015).

7.1 The Formation and Development of the Target Responsibility System of Environmental Protection

In 1986, the Target Responsibility System of Environmental Protection was implemented experimentally in some parts of Gansu province, and expanded to the entire province in 1987. In 1988, this system was implemented, as a trial, in four more provinces, including Shandong, Shanxi, Jiangsu and Zhejiang. In April 1989, in order to promote the comprehensive and thorough development of environmental protection, the Target Responsibility System of Environmental Protection was introduced during the Third National Conference on Environmental Protection. As an administrative management system, its aim was to initiate a liability system of environmental quality for local governments at all levels, in which the scope and allocation of responsibility for various stakeholders in a region or an industrial sector were clearly determined, and target-oriented, quantitative, and institutionalized management methods were employed so as to elevate the national policy of implementing environmental protection as the code of conduct for leaders and cadres at all levels. This is an organically integrated system of authority, responsibility, benefits and obligation. As clearly stipulated in Article 16 of the EP Law (1989), local people's governments at all levels shall be responsible for the environmental quality in their respective jurisdictions and take measures to improve the environmental quality. Thereafter, Target Responsibility System of Environmental Protection was implemented nationwide as an environmental management system (Wang, 2013).

In order to concretely implement the national policy of environmental protection and the sustainable development strategy, to contain the environmental pollution and ecological degradation nationwide by 2000, and to improve the environmental quality of some cities and areas, *the Decisions on Several Issues Concerning Environmental Protection* was issued by the State Council in August 1996. As stipulated, the responsibility system of environmental quality shall be implemented for administrative leadership; local governments at all levels and their main leading cadres and chief officers shall bear the responsibility of environmental protection, according to legislation; and environmental quality of the area shall be regarded as an important indicator of administrative performance evaluation for the leading cadres and chief officers. Local governments at all levels shall regulate major pollutant emissions, to outline the specific objectives and plans for improving environmental quality, and to report to their superiors in government for recording and archiving. Consequently, in order to ensure the implementation of this system, many regional regulations and provisions have been developed in many regions, as shown in Box 7.1.

> **Box 7.1 Major supporting schemes of environmental policy**
>
> - *Trial Measures of Target Responsibility System of Environmental Protection during the Office-Term in Guangdong Province (1991);*
> - *Measures of Implementing the Target Responsibility System of Environmental Protection in Yunnan Province (1994);*
> - *Notice on the Implementation of the Target Responsibility System of Environmental Protection in Guangxi Province (1997);*
> - *Measures of Implementing the Target Responsibility System of Environmental Protection in Guizhou Province (1999);*
> - *Measures of Implementing the Target Responsibility System of Environmental Protection in Hainan Province (1999);*
> - *Schemes of Implementing the Target Responsibility System of Environmental Protection in Jilin Province (2003), etc.*

This system was further consolidated with the promulgation of various relevant laws and regulations, as shown in Table 7.1. For example, as explicitly stipulated in Article 26 of the EP Law (2014), the State shall implement the Target Responsibility System of Environmental Protection and Performance Evaluation System. For the people's governments at or above the county level, the achievement of environmental protection objectives shall be incorporated into the Performance Evaluation System for the competent authorities for environmental protection and their chief officers, at the same level or lower. The results of performance evaluation shall be disclosed to the general public.

7.1 The Formation and Development of the Target Responsibility System of Environmental Protection

Table 7.1 Provisions of the relevant laws and regulations on the target responsibility system of environmental protection and performance evaluation system

Time	Laws and regulations	Relevant provisions
1996	The Decision on Several Issues Concerning Environmental Protection	To implement the responsibility system of environmental quality for administrative leadership
2005	The Decision on Implementing the Scientific Outlook on Development and Strengthening Environmental Protection	To adhere to and perfect the Target Responsibility System of Environmental Protection and Performance Evaluation System for local governments at all levels and exercise the annual target management for the key tasks and objectives of environmental protection
2008	The Law of the People's Republic of China on Water Pollution Prevention and Control	The State shall implement the Target Responsibility System of Water Environmental Protection and Performance Evaluation System
2007	The Law of the People's Republic of China on Energy Conservation	The State shall implement the Target Responsibility System of Energy Conservation and Performance Evaluation System
2011	Opinions Concerning Strengthening the Key Works of Environmental Protection	To formulate the target system of ecological civilization construction and integrate it into the performance evaluation on local governments
2013	The Action Plan for Air Pollution Prevention and Control	In order to be fully responsible for regional air quality, local governments at provincial level shall sign the Liability Statement of the Target of Air Pollution Prevention and Control, and to institute regional enforcement regulations
2014	The Ordinance of Water Supply, Utilizations and Management for South-to-North Water Transfer Project	In order to ensure the water quality of the South-to-North Water Transfer Project, the Target Responsibility System and Performance Evaluation System shall be implemented for the people's governments above the county level
2014	The Law of the People's Republic of China on Environmental Protection	The State shall implement the Target Responsibility System of Environmental Protection and Performance Evaluation System
2015	The Law of the People's Republic of China on Air Pollution Prevention and Control	Local governments at all levels shall be responsible for regional air quality. The competent authority for environmental protection under the State Council shall, in conjunction with the relevant departments under the State Council, evaluate local governments at provincial level for their accomplishment of the target of air quality improvement and the key tasks of air pollution prevention and control. Local governments at provincial level shall formulate the evaluation methods and perform the evaluation on local governments under the provincial level for their accomplishment of the target of air quality improvement and the key tasks of air pollution prevention and control. The results of performance evaluation shall be publicized

7.2 Targets of Environmental Protection and Indicators Setting

In a general manner, the targets of environmental protection are set up in accordance with the directive and normative documents issued by the State Council (such as *the Decisions on Several Issues Concerning Environmental Protection*) and proposed in the FYPs (five-year plans of national economic and social development). Essentially, these documents and FYPs are not laws that have concrete legal effects on non-government bodies. However, from the perspective of internal effectiveness, these "soft laws" do constitute an actual binding force that induces areal effectiveness on the part of government bodies and their staff. In order to fulfill these targets of environmental protection set forth in the directive and normative documents, and in the FYPs, significant legal, policy, and financial support was apportioned and the Performance Evaluation System and the leading cadre evaluation system were implemented (Wang, 2013; Liang & Langbein, 2015). Some targets of environmental protection and the corresponding indicators are displayed in Table 7.2.

7.3 Target Setting

Based upon the national targets, first, the targets of pollution reduction are assigned to each province, according to provincial conditions. Then, the targets of pollution reduction of each province are allocated to various cities and state-owned enterprises under its jurisdiction, according to individual conditions. In order to create a buffer zone for any unachieved targets from lower-performing governments, higher-performing governments will usually assign more stringent requirements and higher targets to these lower-performing governments. However, less developed provinces (counties and districts) would be assigned less stringent requirements and lower targets, to clearly reflect the core objective of reducing the economic disparity between the developed provinces in the coastal regions and the less developed provinces in western China (Wang, 2013). For example, the target of $PM_{2.5}$ concentration reduction for Beijing was slightly higher than the national target. However, various targets were assigned to different counties and districts according to regional needs and local conditions, as shown in Table 7.3.

7.4 Performance Evaluation

Performance evaluation plays a significant role in promoting the Target Responsibility System by compelling the local governments and their officers to carry out the environmental protection work actively and effectively and identifying the problems

7.4 Performance Evaluation

Table 7.2 Targets of environmental protection set forth in the normative documents and the FYPs

Documents and the FYPs	Targets of environmental protection and indicators of performance evaluation
The Decision on Several Issues Concerning Environmental Protection (1996)	By 2000, the deteriorating trend of environmental pollution and ecological degradation shall be contained, and the environmental quality in some cities and regions shall be greatly improved
The 10th FYP (2001–2005)[a]	By 2005, the status of environmental pollution will be alleviated; the trend of ecological degradation will be contained; urban-rural environmental quality, especially for large and medium-sized cities and key areas, will be improved; and the laws, policies and management system of environmental protection will be improved and enhanced to well adapt to the socialist market economic system By 2005, the emissions of SO_2 in the acid rain control zone and SO_2 control zone shall be reduced by 20% compared to that in 2000
The 11th FYP (2006–2010)	By 2010, the emissions of SO_2 and COD shall be reduced by 10% compared to that in 2005
The 12th FYP (2011–2015)	8% reduction for SO_2 and COD; 10% reduction for NH_3-N and NO_x; and Focusing on cutting heavy-metal pollution from industry
The 13th FYP (2016–2020)	By 2020, the emissions of COD and NH_3–N shall be reduced by 10% compared to that in 2015; and the emissions of SO_2 and NO_x shall be reduced by 15% compared to that in 2015
The Action Plan for Air Pollution Prevention and Control (2013)	By 2017, urban PM_{10} concentration, nationwide, shall be reduced by 10% compared to that in 2012; and the $PM_{2.5}$ concentrations in the Beijing-Tianjin-Hebei Area, the Yangtze River Delta Area and the Pearl River Delta Area shall be reduced by 25%, 20% and 15%, respectively
The Action Plan for Water Pollution Prevention and Control (2015)	By 2020, China's water environment quality shall be gradually improved; the percentage of severely polluted water bodies shall be greatly reduced and the quality of drinking water shall be elevated By 2020, for more than 70% of the watershed of seven major rivers, such as the Yangtze River, the Yellow River, etc., the water quality shall be level III or above; and the foul water body in urban built-up areas shall be controlled to no exceeding 10% By 2030, for more than 75% of the watershed of seven major rivers, such as the Yangtze River, the Yellow River, etc., the water quality shall be level III or above; and there shall be no foul water body in urban built-up areas

(continued)

Table 7.2 (continued)

Documents and the FYPs	Targets of environmental protection and indicators of performance evaluation
The Action Plan for Soil Pollution Prevention and Control (2016)	By 2020, seriously deteriorating trend of soil contamination shall be controlled to keep stable soil quality; safe utilization of contaminated cultivated lands shall be around 90% and safe utilization of other contaminated lands shall be over 90% By 2030, safe utilization of contaminated cultivated lands shall be 95% and safe utilization of other contaminated lands shall be over 95% To gradually improve soil quality across the country by mid-21st century
Three-Year (2018—2020) Action Plan for Winning the Blue-Sky War (2018)[b]	By 2020, the emissions of SO_2 and NO_x shall be reduced by more than 15% compared to that in 2015 For $PM_{2.5}$ nonattainment cities, $PM_{2.5}$ concentration shall be reduced by at least 18% compared to that in 2015 For city at county level or above, the ratio of days with good air quality shall be 80%, annually, and the percentage of heavily polluted days shall be decreased by more than 25% compared to that in 2015 Provinces have completed the targets of the 13th FYP ahead of schedule shall maintain their achievements. Those have not yet completed the targets of the 13th FYP shall ensure the binding targets of the 13th FYP will be fully achieved

[a]Usually, the contents of energy saving, emissions reduction, climate change, resource saving, natural ecosystems, biodiversity, and so on, are illustrated in the chapter of environmental protection of the FYP. Here, the target of emissions reduction was selected as an example
[b]*Three-Year (2018–2020) Action Plan for Winning the Blue-Sky War* is seen as the second phase of the Action Plan for Air Pollution (2013)

and deficiencies in the implementation process. Therefore, a set of scientific and reasonable indicators for the Performance Evaluation System can be quite essential and critical to successfully implement the Target Responsibility System of Environmental Protection. Some evaluation guidelines and methods for the Target Responsibility System of Environmental Protection are listed in Table 7.4.

7.5 Shared Responsibility of Ecological and Environmental Protection for Government and Party

The notion of the "Five-in-One" Strategy is the overall plan of promoting socialism with Chinese characteristics, proposed during the 18th NCCPC, in which ecological civilization construction, economic construction, political construction, cultural construction and social construction shall be thoroughly integrated to build a sustainable and beautiful China. Since the 18th NCCPC, it has been repeatedly stressed upon by the General Secretary Xi Jinping to improve the responsibility system of

7.5 Shared Responsibility of Ecological and Environmental Protection for Government and Party

Table 7.3 Target setting of PM$_{2.5}$ concentration in Beijing

		PM$_{2.5}$ concentration reduction by 2017, compared to 2012 level (%)	Annual average PM$_{2.5}$ concentration ($\mu g/m^3$)
National target		≈25	≈60
Provincial target		≥25	≈60
Target at county and district level	Huairou, Miyun and Yanqing	≥25	≈50
	Shunyi, Changping and Pinggu	≥25	≈55
	Dongcheng, Xicheng, Chaoyang, Haidian, Fengtai and Shijingshan	≥30	≈60
	Mentougou, Fangshan, Tongzhou, Daxing and Beijing Economic and Technology Development Area	≥30	≈65

Source Beijing Clean Air Action Plan (2013–2017).
Note The main targets of Beijing's Clean Air Action Plan were split into 84 specific tasks to involve more than 30 responsible bodies, including local districts, Beijing Development and Reform Commission, Beijing Municipal Commission of Transport, Beijing Transport Management Bureau, Beijing Environment Protection Bureau (BEPB) and other related authorities

ecological and environmental protection and to prohibit any destruction of ecological civilization with a firm attitude and decisive measures.

As proposed in the Bulletin of the Third Plenary Session of the 18th CCCPC (November 9–12, 2013), the lifetime liability system shall be founded for ecological and environmental damage. Further, as emphasized in the Bulletin of the Fourth Plenary Session of the 18th CCCPC (October 20–23, 2014), a more stringent legal system shall be applied to ecological and environmental protection, and the lifetime liability system shall be established for major decision-making, as well as the mechanism of responsibility investigation, according to the requirements of comprehensive advancement of the rule of law. As clearly stipulated in *the Opinions on Accelerating the Advancement of Ecological Civilization Construction*, promulgated by the CCCPC and the State Council, jointly, in May 2015, liability for severe ecological and environmental damage shall be pursued for lifetime.

In August 2015, *the Measures of Pursuing the Liability of Party and Government Leading Cadres for Ecological and Environmental Damage (on Trial)* was promulgated through the collaborative efforts of the NDRC, MOF, MLR, MEP, MHURD, MWR, MOA, and SFA, led by the Organization Department of the CCCPC and the

Table 7.4 Evaluation guidelines and methods for the target responsibility system of environmental protection

Evaluation guidelines	Evaluation method	Significance
The Measures of Performance Evaluation on Total Emission Reduction of Major Pollutants during the 12th FYP	Qualitative and quantitative indicators Self-reporting and on-site inspection Scoring system	The results shall be an important reference to cadre's annual reviews for bonuses, prizes and promotion For a region with poor results, the MEP shall suspend the review and approval process of EIA documents for all new projects with major pollutants in the region According to the regulations of performance management for emission reduction, the Committee of Supervisory at the same level along with the competent authority for environmental protection at superior level, shall take actions against the leading cadre, such as publicized criticism, reprimand, admonition, verbal warning, etc.
The Measures of Performance Evaluation on the Implementation of the Action Plan for Air Pollution Prevention and Control (on Trial)	Self-reporting and on-site inspection Regular inspection and environmental protection inspections by the MEP Annual evaluation and final assessment Scoring system	

Ministry of Supervision, to further solidify the implementation of the lifetime liability system for ecological and environmental damage, and to explicitly demonstrate the determination to pursue leading cadres' liability for ecological and environmental damage from both the party and the government. The introduction and implementation of the abovementioned Measures have enhanced the ecological civilization awareness of the leading cadres at all levels, generated far-reaching effects on the leading cadres' behavior for ecological and environmental protection, and promoted the advancement of ecological civilization construction.

7.6 The Audit of Natural Resources Assets for Leading Cadres While Leaving Office

As set forth in the Bulletin of the Third Plenary Session of the 18th CCCPC, the audit of natural resources assets shall be imposed on leading cadres when they leave office, as the essential reform measures of enhancing the ecological civilization construction. In November 2015, the General Office of the CCCPC and the General Office of the State Council jointly promulgated the *Pilot Program of Deploying the Audit of Natural Resources Assets for Leading Cadres while Leaving Office*, an important component of the ecological civilization system to significantly promote the ecological civilization construction.

Key areas covered by the audit include land resources, water resources, forest resources, environmental governance of mining, air pollution prevention and control, and other fields. The audit is primarily focused on the performance evaluation of natural resources assets management and ecological and environmental protection during the term of office for the leading cadres, so as to determine the scope of their responsibilities. The purposes of audit are, on the one hand, to reveal the prominent problems between environmental protection and natural resources exploitation and utilization, and the potential safety risks for natural resources and ecological environment; and, on the other hand, to realize the Target Responsibility System and to enhance the lifetime liability system so as to guide the leading cadres toward the correct concept of administrative accomplishment.

References

Liang, J., & Langbein, L. (2015). Performance management, high-powered incentives, and environmental policies in China. *International Public Management Journal, 18*(3), 346–385. https://doi.org/10.1080/10967494.2015.1043167. [2019-08-05].

Lo, K. (2015). How authoritarian is the environmental governance of China? *Environmental Science & Policy, 54,* 152–159. https://doi.org/10.1016/j.envsci.2015.06.001. [2019-08-05].

Wang, A. L. (2013). The search for sustainable legitimacy: Environmental law and bureaucracy in China. *Harvard Environmental Law Review, 37,* 365–440. https://doi.org/10.2139/ssrn.2128167. [2019-08-05].

Chapter 8
Centralized Pollution Control System

The Centralized Pollution Control System can be described as the application of centralized prevention and control measures to similar pollution sources within an area, based on the premise that the obligation of the polluter to prevent and treat pollution shall never be less, and in which the effectiveness and economies of scale can be greatly enhanced while improving the environmental quality. In recent years, the Centralized Pollution Control System was mainly employed in the following categories, including central heating, centralized treatment of wastewater, and centralized disposal of solid waste and hazardous waste.

In February 1981, it was pointed out in *the Decisions on Strengthening the Environmental Protection Work during the Period of National Economy Adjustment* promulgated by the State Council that in urban planning and construction centralized heating and heating for contiguous areas shall be heavily promoted to gradually phase out outdated distributed heating systems, such as the one-boiler-to-one-unit mode. In October 1984, as clearly stated in *the Technical Guidance on Pollution Prevention and Control of Smoke from Coal-Burning* issued by the EPC, distributed heating systems shall be replaced by centralized heating systems in the overall planning of renovation of old cities and the construction of new ones. During the Third National Conference on Environmental Protection held in April 1989, as explicitly specified by the EPC, in the context of China's realities, the integration of centralized pollution control and distributed pollution control shall be adopted for pollution governance in which centralized pollution control shall play a leading role in future development. Therefore, the Centralized Pollution Control System was formally confirmed.

In Article 51 of the EP Law (2014), it was stipulated that people's governments at all levels shall be responsible for the overall urban and rural planning to construct the wastewater treatment facilities and supporting pipeline networks, the sanitation facilities for solid waste collection, transportation and disposal, the centralized sites and facilities for hazardous waste disposal, and other public facilities for environmental protection, as well as to ensure the normal operation of these facilities. At present in China, the main targets of centralized pollution control are wastewater, air pollution and solid waste.

8.1 Centralized Heating in Urban Areas

In the circumstances in which many laws and regulations have been enacted to greatly promote energy conservation and emission reduction, the centralized heating system, as a means of energy conservation and less environmental pollution, has gradually become the key mode of heating supply in urban Chinese areas. For industrial complexes, urban areas, and congregated residential compounds, the centralized heating system is a form of heating supply to urban residents and surrounding industries for domestic livelihood and industrial production.

Due to the fact that China has the highest population in the world and a higher population density in most of its cities, the centralized heating system shall be more appropriate for urban areas in northern China. Currently, the centralized heating system is widely adopted in most of the northern provinces, including Heilongjiang, Jilin, Liaoning, Xinjiang, Qinghai, Gansu, Ningxia, Beijing, Tianjin, Hebei, Shanxi, and Neimenggu, and in parts of Shaanxi, Henan, and Shandong.

In recent years, the total area and amount of heating supplied by centralized heating systems in China have been continuously increasing, as shown in Fig. 8.1. At present, the most common type of centralized heating system is the Combined Heat and Power (CHP) Cogeneration System, supplemented by the centralized boiler room and others. Coal is the major energy source of centralized heating systems, while natural gas is used as an energy source in only a few cities, such as Beijing, Shanghai, and Urumqi. In order to fulfill the norms of energy conservation, emission reduction, and sustainable development for national economic and social development, it is necessary to intensely promote the construction and development of centralized heating

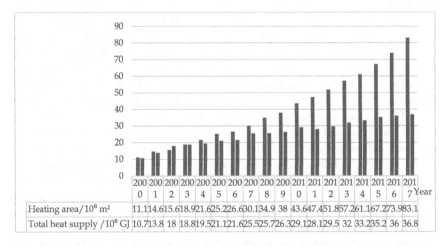

Fig. 8.1 Areas and amount of heating supply of centralized heating system during winter time in China. *Data Source* The Ministry of Housing and Urban-Rural Development of the People's Republic of China, *the Statistical Yearbook of Urban and Rural Development.* and *the Bulletin of Urban and Rural Construction Statistics.*

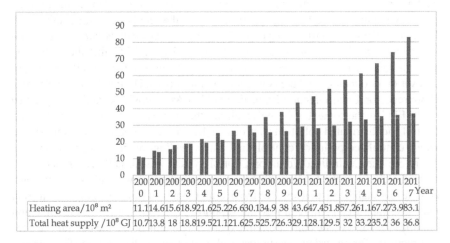

systems. With such key features as high efficiency of heat generation, energy conservation, and lower environmental pollution, the CHP Cogeneration System shall be the major development direction of urban heating supply technologies, while the large-scale centralized boiler room will serve as the main supplement to the centralized heating system, and the distributed boiler room shall be gradually phased out. In addition, to progressively reduce the use of coal, natural gas shall be the major energy source for urban heating supply, supplemented by other clean energy sources.

8.2 Centralized Pollution Control of Wastewater

The distinct features of wastewater treatment development in China can be summarized as follows: ① the volume of wastewater discharged was continuously increased; ② the development and capacity building of wastewater treatment in cities and towns was significantly enhanced; ③ the capacity building of wastewater treatment in rural areas shall be considerably elevated.

(1) Continuously increasing discharged wastewater

In the 21st century, along with rapid industrialization and urbanization, the volume of wastewater discharged was significantly increased in China, as shown in Fig. 8.2. For domestic wastewater, the volume was steadily increased, obviously, as the ratio to total wastewater was continuously elevated to reach more than 70% in 2015. For industrial wastewater, the volume showed an increase from 2000 to 2007, and then began to decline ever since, which might be attributed to the promotion of cleaner production, circular economy, and reutilization of reclaimed water for industrial sectors.

(2) Fast development and capacity building of wastewater treatment in cities and towns

In recent years, the State has seriously promoted energy conservation and emission reduction to prevent and control water pollution. In addition to the enactment of relevant policies and regulations, both central and local governments have been continuously increasing their investment in the infrastructure construction of urban wastewater treatment facilities, along with the introduction of a market mechanism and standards system. By the end of 2007, there were 883 wastewater treatment plants (WWTPs) in cities and towns, nationwide, with a processing capacity of 7.138×10^7 m^3/day, accounting for 62.80%. According to the *Notification on the Construction and Operation of Wastewater Treatment Facilities in Cities and Towns in the First Half of 2017*, published by the MHURD in November 2017, there were 4063 WWTPs built and operated in cities and towns, nationwide, with a processing capacity of 1.78×10^8 m^3/day by June 2017; there were 2327 WWTPs built and operated in cities with a processing capacity of 1.48×10^8 m^3/day, and 1736 WWTPs built and operated in 1470 counties (accounting for 94.2% of the total number of counties) with a processing capacity of 3.1×10^7 m^3/day. Furthermore, there were 570

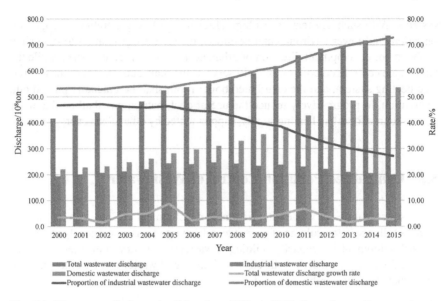

Fig. 8.2 Wastewater discharge in China from 2000 to 2015. *Data Source* Data on processing capacity of WWTP were extracted from the National Bureau of Statistics. The number of wastewater treatment plant (WWTP) was acquired from *the Statistical Bulletin of Urban and Rural Construction* issued by the Ministry of Housing and Urban-rural Development of the People's Republic of China.

WWTPs built and operated in 36 key cities with a processing capacity of 6.5×10^6 m^3/day. Noticeably, in China, the number and processing capacity of WWTPs, and the rate of treated wastewater have been considerably increased and enhanced in the new millennium, as shown in Fig. 8.3.

(3) Capacity building of wastewater treatment in rural areas shall be elevated

In undeveloped areas located in Midwest China and rural areas, due to lower economic strength, limited financial support, and dispersed population, the construction and operation of wastewater treatment facilities are still far behind. According to *the Bulletin of Urban and Rural Construction Statistics of 2014*, publicized by the MHURD, there were 7.63×10^8 rural people allocated among 2.7×10^6 natural villages and 5.4×10^5 administrative villages, in which wastewater was treated in only 9.98% of the administrative villages. Furthermore, the rate of treated wastewater at the village level was only 11.4% in 2015. Thus, the ubiquitous problems of wastewater treatment facilities in rural areas can be summarized as follows: insufficient pipeline network, low collection rate of wastewater, and outdated treatment techniques resulting in most of the wastewater being discharged arbitrarily without any treatment.

8.2 Centralized Pollution Control of Wastewater

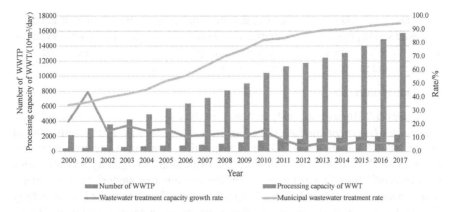

Fig. 8.3 Wastewater treatment facilities and capacity in China from 2000 to 2017. *Data Source* Processing capacity of WWTPs, Nation Bureau of Statistics. The number of wastewater treatment plant (WWTP): from *the Bulletin of Urban and Rural Construction Statistics* issued by the Ministry of Housing and Urban-rural Development of the People's Republic of China.

As explicitly specified in *the Action Plan for Water Pollution Prevention and Control* (the Action Plan for Water Pollution) enacted by the State Council in April 2015, there is an urgent need to speed up the comprehensive governance of the rural environment, as shown in Box 8.1.

Box 8.1 Major tasks of comprehensive governance of rural environment in the Action Plan for Water Pollution

- *To implement the integrated planning, construction, management and supervision on wastewater treatment in villages, by adopting the administrative area of a county as the base unit.*
- *For areas with feasible conditions, the services of wastewater treatment shall be extended and provided to the villages from urban wastewater treatment facilities.*
- *To promote governance through incentives and rewards, to implement cleaner engineering in villages, to dredge and clean out river channels, and to facilitate rural environment governance of contiguous villages.*
- *By 2020, the comprehensive governance of rural environment shall be accomplished for 1.30×10^5 administrative villages.*

Several actions were proposed in *the Opinions Concerning Deeply Promoting the Construction of New Type of Urbanization* disseminated by the State Council in February 2016, such as the promotion of rural living environment governance; enhancement of the construction of collection and treatment facilities for solid waste and wastewater in villages; and the facilitation of integrated planning for wastewater governance in villages. On the demand side, in villages, there is a vast need for domestic waste governance, domestic wastewater governance, pollution prevention and control for area sources, environmental governance of watersheds and river channels, and resource utilization of livestock and poultry breeding waste.

(4) Future prospects

In December 2016, as noticeably indicated in *the Construction Plan for the Urban Wastewater Treatment and Recycling Facilities during the 13th FYP*, jointly enacted by the NDRC and the MHRUD, the construction of urban wastewater treatment facilities shall be transformed from "scale expansion" to "quality and efficiency advancement", from "focusing on wastewater and ignoring sludge" to "equal attention to both wastewater and sludge", from "treatment" to "recycling", so as to fully elevate the capacity and services of urban wastewater treatment facilities in China.

8.3 Centralized Pollution Control of Municipal Solid Waste

In China, the amount of municipal solid waste (MSW) has been increased rapidly during the past 40 years, with an annual average growth rate of about 5.5% as displayed in Fig. 8.4. Urbanization, industrialization, and rapid population growth are the three key driving forces behind the extensive generation and rapid increase in the amount of MSW in China (Zhang, Tan, & Gersberg, 2010). The major components

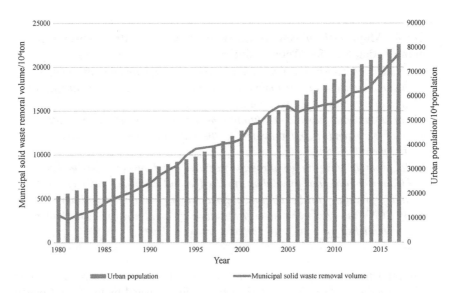

Fig. 8.4 Municipal solid waste generation and urban population in China from 1980–2016. *Data Source* Nation Bureau of Statistics.

of MSW in China are food waste (around 60% from 1989 to 2014), ash (decreased from 27.2% in 1989 to 2.5% in 2014), plastic (varied from 9.4% in 2007 to 7.1% in 2010, and decreased ever since), and paper (fluctuated from 6 to 10% annually during the last decade) (Gu et al., 2017). Currently, there are several key methods for MSW disposal, including landfill, incineration, and so forth.

8.3.1 Landfill

Landfill has been the primary method of solid waste disposal because it is cheaper and more convenient in comparison to other methods of solid waste disposal. In China, the construction of sanitary landfills was initiated in the late 1980s. Currently, many landfill sites have reached their designed capacity, and it is becoming far more difficult to find appropriate sites to construct new landfills, both in urban areas and in the countryside. During the 13th FYP (2016–2020), the number of landfills will peak around 2,400, and subsequently reduced slightly, stabilizing at about 2,000–2,200, as shown in Table 8.1.

8.3.2 Incineration

Incineration is the process of burning solid waste in a controlled manner. Incineration is a feasible option for large cities where space for landfills may be limited and the siting of landfills outside of city jurisdictions may be difficult. MSW incineration facilities were first introduced into China in the late 1980s (Li, Zhang, Li, & Zhi, 2016). The number of MSW incineration facilities has shown a steady increase during the past 30 years. Most of the incinerators are designed to capture heat for electricity generation while burning solid waste, whereas some incinerators are only used to burn solid waste. The first waste incineration power plant was constructed in Shenzhen in 1985 (Xu et al., 2015). According to previous research, waste incineration power was the second biggest industry in biomass power generation in China, second only to power generated from agricultural and forest waste (He & Lin, 2019).

With the growing difficulties of constructing new landfill sites and incineration facilities due to land scarcity and public concerns in China, more and more alternatives of solid waste management are being explored, including waste separation and recycling.

Table 8.1 Solid waste treatment unit and disposal capacity in China

Year	Number of harmless treatment plant/unit			Harmless treatment capacity/(10^4 ton/day)			Volume of harmlessly treated/10^4 ton		
	Landfill	Incinerate	Other	Landfill	Incinerate	Other	Landfill	Incinerate	Other
2006	433	72	24	22.1	4.0	1.0	6770.2	1142.4	321.7
2007	497	68	20	23.3	4.5	0.8	8086.0	1446	279.0
2008	605	78	20	28.5	5.4	0.6	9161.4	1615.4	214.2
2009	710	103	23	31.5	7.4	0.8	9945.2	2122.1	206.3
2010	919	119	23	30.2	9.0	0.8	11134.5	2432.7	218.0
2011	1195	130	33	39.4	10.1	1.7	12317.6	2801.6	494.8
2012	1335	167	47	42.5	13.2	1.7	13791.4	3876.4	512.8
2013	1522	197	37	46.0	17.0	1.5	14398.5	4931.7	363.1
2014	1733	222	66	48.2	20.0	1.8	15005.5	5674.1	480.8
2015	1748	257	72	50.2	23.5	2.0	16171.3	6577.4	524.3
2016	1840	299	74	5	27.8	2.2	16779.5	7956.2	617.8
2017	1852	352	109	52.6	33.1	2.8	17125.4	9321.2	727.3

Data Source Nation Bureau of Statistics. *The Statistical Yearbook of Urban and Rural Development* issued by the Ministry of Housing and Urban-rural Development of the People's Republic of China.

References

Gu, B., Jiang, S., Wang, H., Wang, Z., Jia, R., Yang, J., et al. (2017). Characterization, quantification and management of China's municipal solid waste in spatiotemporal distributions: A review. *Waste Management, 61,* 67–77. https://doi.org/10.1016/j.wasman.2016.11.039. [2019-08-26].

He, J., & Lin, B. (2019). Assessment of waste incineration power with considerations of subsidies and emissions in China. *Energy Policy, 126,* 190–199. https://doi.org/10.1016/j.enpol.2018.11.025. [2019-08-26].

Li, X., Zhang, C., Li, Y., & Zhi, Q. (2016). The status of municipal solid waste incineration (MSWI) in China and its clean development. *Energy Procedia, 104,* 498–503. https://doi.org/10.1016/j.egypro.2016.12.084. [2019-08-26].

Xu, Y., Chan, A. P. C., Xia, B., Qian, Q. K., Liu, Y., & Peng, Y. (2015). Critical risk factors affecting the implementation of PPP waste-to-energy projects in China. *Applied Energy, 158,* 403–411. https://doi.org/10.1016/j.apenergy.2015.08.043. [2019-08-26].

Zhang, D. Q., Tan, S. K., & Gersberg, R. M. (2010). Municipal solid waste management in China: Status, problems and challenges. *Journal of Environmental Management, 91*(8), 1623–1633. https://doi.org/10.1016/j.jenvman.2010.03.012. [2019-08-26].

Chapter 9
Emission Reporting, Registration and Permit System

There are two key parts of the Emission Reporting, Registration and Permit System ("the Emission Permit System" in short), namely, emission reporting and registration, and emission permit, respectively. First, the polluter shall report to and register with the competent authority for environmental protection for its pollution conditions, including pollutant(s) emission facilities, pollutant(s) treatment facilities, and the types, quantities and emission concentrations of pollutants under normal operating conditions, as per the relevant regulations. Second, according to the application of the polluter, the competent authority for environmental protection shall issue an emission permit to the polluter specifying the allowable emission terms for the polluter under normal operating and production conditions. As clearly defined in Sect. 6.2.1, whichever entity directly discharges pollutants (including air pollutants, water pollutants, noise, and solid wastes) into the environment is a polluter; they can be categorized as key pollutants polluters and general pollutants polluters. The Emission Permit System is a legal administrative management system aiming to improve the environmental quality by regulating the amount and volume of pollutants discharged, and the destination of discharge pollutants, based on the Emission Cap System, which will be described comprehensively in Chap. 10.

9.1 The Formation and Development of the Emission Permit System

The Emission Permit System is an important environmental management system in China. The pilot program of the Emission Permit System was initiated by the NEPA in 1987, in which 18 cities, including Shanghai, Hangzhou, and others, were selected for implementation of the pilot program. In March 1988, *the Interim Measures of Management for Water Pollutants Emission Permit* was promulgated by the NEPA, with the aim of controlling the total amount and volume of water pollutants discharged through issuing of water pollutant emission permits after reporting. In April 1989, it was proposed to promote the implementation of the Emission Permit System, nationwide, during the Third National Conference on Environmental Protection. In July 1989, the *Regulations on the Implementation of the Law of the People's Republic of China on Water Pollution Prevention and Control* was publicized by the NEPA to explicitly stipulate in Article 9 that any entity whose pollutant emission is in compliance with national and local pollutant emission standards and does not exceed the pollutant emission cap for that particular enterprise will be granted an emission permit by the competent authority for environmental protection. As further specified in Article 27 of the EP Law (1989), any entity with pollutant emission shall report to and register with the competent authority for environmental protection, in accordance with the relevant regulations.

As specified in the *Decisions on further Enhancing the Environmental Protection Work*, promulgated by the State Council in December 1990, the competent authority for environmental protection at all levels shall gradually implement the Emission Cap System and the Emission Permit System. As regulated in Article 8 of the *Regulations on the Implementation of the Law of the People's Republic of China on Air Pollution Prevention and Control*, enacted by the State Council in May 1991, any entity with pollutant emission shall report to and register with the local competent authority for environmental protection. Therefore, in 1991, 16 cities, including Shanghai, Tianjin, Taiyuan, Guangzhou, and Shenyang, were selected by the NEPA to implement the pilot program of air pollutants emission permits.

According to the *Regulations of Management for the Pollutant Emission Reporting and Registration* issued by the NEPA in August 1992, pollutant emission reporting and registration shall be carried out throughout the nation, and the work of pollutant emission reporting and registration for cities directly under the provincial government and above shall be completed by the end of 1995. Moreover, the emission reporting and registration of industrial solid waste and environmental noise pollution were stipulated in the *Law of the People's Republic of China on Solid Waste Pollution Prevention and Control* (decreed in October 1995) and the *Law of the People's Republic of China on Environmental Noise Pollution Prevention and Control* (decreed in October 1996), respectively. In January 1997, the specific requirements for the comprehensive implementation of emission reporting and registration, nationwide, were set forth in the *Notice on Comprehensive Implementation of Emission Reporting and Registration*, issued by the NEPA.

As stipulated in Article 10 of *the Regulations on the Implementation of the Law of the People's Republic of China on Water Pollution Prevention and Control* (2000 Revision), promulgated in March 2000, the water pollutant emission permit shall be granted by the local competent authority for environmental protection, according to the schemes of the emission cap. Up until this point, the Emission Permit System of Water Pollutants was officially established through administrative regulations. In addition, as specified in Article 15 of the Air Pollution Law (2000 Revision), promulgated in April 2000, within the major air pollutants emission control areas, the air pollutant emission permit shall be granted by the local competent authority for environmental protection, according to the conditions and procedures stipulated by the State Council. Subsequently, the Emission Permit System was implemented

9.1 The Formation and Development of the Emission Permit System

under the Emission Cap System. As the implementation of the Emission Cap System for major pollutants since the 9th FYP (1996–2000), the importance of the Emission Permit System for China's environmental protection work has been continuously increasing.

In February 2008, it was clearly stipulated in Article 20 of the Water Pollution Law (2008 Revision) that the Emission Permit System shall be implemented throughout the nation, signaling that the development of China's Emission Permit System had been elevated to a substantive stage. Further, it was stated in the *Decision on Several Major Issues Concerning the Comprehensively Deepening Reform*, promulgated by the CCCPC in November 2013, that the Emission Permit System shall be further improved and the Emission Cap System shall be implemented for enterprise. After years of development and implementation, though the principal regulations on the Emission Permit System have been enshrined in various revised laws, including the Air Pollution Law (2000 Revision), the Water Pollution Law (2008 Revision) and the Environmental Protection Law (2014 Revision), there were still many problems with the Emission Permit System, such as low operability, no sufficient linkage with other pollution sources management systems, and so on.

One of the requirements of the *Action Plan for Air Pollution Prevention and Control* (the Action Plan for Air Pollution), enacted by the State Council in September 2013, was perfecting the Emission Cap System and the Emission Permit System. As clearly expressed in the *Action Plan for Water Pollution Prevention and Control* (the Action Plan for Water Pollution), issued by the State Council in April 2015, the Emission Permit System shall be fully implemented and emission permits for all pollution sources shall be completely issued by the end of 2017. The Emission Permit System was initially deployed from key pollution sources and a pilot program of compensated and tradable emission allowances, and gradually extended to all stationary sources. Article 35 of the *General Planning for Ecological Civilization System Reform*, issued by the State Council in September 2015, clearly specified the perfecting of the emission permit system and the establishment of an unified and fair emission permit system to cover all stationary sources, as soon as possible. In addition, it was further stated in Article 54 that laws and regulations to improve the emission permit system were required. These have generated a significant influence on the reform of the environmental management system, especially for the implementation of integrated environmental management, and lawful and effective supervision, monitoring and enforcement of environmental quality control.

Since the 1980s, the Emission Permit System has been implemented in 27 provinces and municipalities, and more than 240,000 emission permits have been granted to polluters. In November 2016, *the Implementation Measures of Controlling the Emission Permit System to Control Pollutant Emission* was released by the General Office of the State Council as the top-layer design of the emission permit system to institute a set of regulations and rules, and to set up a structure for the implementation of the Emission Trading System.

9.2 Implementation Process of the Emission Reporting, Registration, and Permit System

The Emission Permits System is mainly implemented through four stages: reporting and registering pollutant emission, verifying pollutant emissions, issuing the emission permit, and supervision and management of the emission permit.

9.2.1 Reporting and Registering Pollutant Emission

To report and register pollutant emission is the most fundamental work of environmental management and the essential part of the Emission Permit System, in which the emission limits and emission charges are determined. In addition, it is an important means for the competent authority for environmental protection to institute environmental planning, stipulate environmental standards, supervise and monitor pollutant emission, and perform dynamic management for pollution prevention and control within their jurisdiction. The main content of emission registration includes the basic information about the polluter; the geographical position and layout of the factory; the consumption of raw materials, resources and energy; technological process; the type and total amount and volume of pollutants discharged, and their destination; and the construction and operation of the pollution treatment facilities.

9.2.2 Verifying Pollutant Emission

The verification of the pollutant emission is the core work of the Emission Permit System, in which the competent authority for environmental protection shall review the registration form of the polluter and verify the accuracy of the contents of the registration so as to determine the emission conditions, including the allowable amount and volume of pollutant to be discharged, the highest emission concentration, and so on.

9.2.3 Issuing the Emission Permit

The emission permit shall specify the emission conditions, such as the allowable amount and volume of the pollutant to be discharged, the outlet for discharging the pollutant, the pattern of emission, the highest emission concentration, and so forth, so as to regulate the behavior of the polluter. For those polluters that can meet the emission conditions specified in the emission permit, an emission permit shall be granted. Otherwise, further emission reduction on the part of the polluter shall be required.

9.2.4 Supervision and Management for the Emission Permit

Supervision and management of the emission permit, including self-reporting, supervision, monitoring, inspection, auditing, evaluation, and renewal, serve as important follow-up actions and are very critical for ensuring the effectiveness of the emission permit. Once the emission permit is granted, it is necessary to supervise and monitor the actual pollutant emission to see if it is in compliance with the emission permit.

9.3 Essentials of the Emission Permit System

9.3.1 Who Shall Apply for the Emission Permit

The emission permit can be used for seven types of pollution activities: ① discharging industrial waste gas, or toxic and harmful air pollutant(s) prescribed by the State; ② discharging industrial wastewater and medical wastewater to water bodies; ③ centralized heating facilities; ④ large-scale livestock and poultry farms; ⑤ urban or industrial centralized wastewater treatment plant; ⑥ centralized solid waste disposal facilities and hazardous waste disposal facilities; ⑦ other specific activities required by legislation.

9.3.2 An Integrated Emission Permit

An integrated emission permit shall comprehensively describe the emission conditions of all pollutants (e.g., air pollutants, water pollutants, noise, and solid wastes) discharged by the polluter sited at the specific location. The first period of validity of an integrated emission permit is three years, upon expiration of which, the emission permit holder can apply for a renewal for another five years, subject to review and approval from the competent authority for environmental protection.

9.3.3 Procedure to Apply for the Emission Permit

The polluter is required to report to and register with the competent authority for environmental protection regarding the emission conditions, including pollutant(s) discharged, types, amount, and volume of emission; production process where pollutants are generated and discharged; and pollution treatment facilities. The competent authority for environmental protection shall review the application documents, undertake the on-site inspection for verification, if necessary, and then issue the emission permit to the polluter according to the relevant laws and regulations.

9.3.4 Conditions for Obtaining an Emission Permit

In order to be granted a emission permit, the following conditions shall be fulfilled: pollutant emission shall be in compliance with relevant emission standards; the discharging method and destination of pollutant discharged shall conform with the requirements of ecological red line and environmental function zoning; the EIA documents of the construction project shall be approved by or archived at the competent authority for environmental protection; the facilities of pollution prevention and control, and the capacity of pollutants treatment shall be incompliance with relevant national or local standards. In addition, in the case of a polluter discharging key pollutants, a monitoring system shall be installed and operated in accordance with relevant national regulations and monitoring standards, and the setup of the pollutant outlets shall be in line with national or local requirements. Prior to the issuance of the emission permit, all relevant information shall be disclosed by the competent authority for environmental protection, in advance.

9.3.5 Evaluate and Determine the Allowable Emission Based on EIA Results

The annual allowable amount and volume of pollutant to be discharged by the polluter shall be evaluated and determined on the basis of EIA results. Several specific provisions can be summarized as follows: the annual allowable emission from polluter shall not exceed the amount determined by relevant national or local standards; in principle, the maximum daily allowable emission shall not exceed twice the daily average of annual allowable emission under normal operating conditions; for nonattainment areas, more stringent emission concentration limits and emission cap shall be applied; the term of the emission permit shall be valid for no more than five years, which can be renewed under the conditions of complying with all stipulations; for key pollutants, in case of noncompliance with the emission permit, the emission permit shall not be renewed.

9.3.6 Self-reporting and Self-monitoring

The polluter shall report its actual pollution emission and be responsible for the authenticity, accuracy and completeness of the report, as well as disclose all information to the public. All emission conditions shall be in compliance with that prescribed in the emission permit. In case of noncompliance with the emission permit, the polluter shall report and inform the competent authority for environmental protection, immediately. In addition, the polluter is also required to carry out self-monitoring activity by adopting legitimate monitoring schemes to ensure the normal operations of pollution prevention and control facilities. Original monitoring data must be saved and archived to establish an accurate and complete account of environmental management. Polluters obliged to install online monitoring devices shall share all online monitoring data with the competent authority for environmental protection.

9.3.7 *Information Disclosure and Public Supervision*

The monitoring data submitted by the polluter will be publicized on the website of the Permitting Management Information System, along with the enforcement information and a list of polluters who are not in compliance with the emission permits, provided by the competent authority for environmental protection. Noncompliance will result in serious degra-dation of the credit rating of polluter's emission behaviors, which will be displayed on the National Enterprise Credit Information Publicity System. The public are encouraged to report noncompliance, such as discharging pollutants not in accordance with the emission permit or even without the emission permits.

9.3.8 *Enforcement and Responsibilities*

The online monitoring data of polluters can be used as a legal basis for the competent authority for environmental protection to fulfill its legal duties, including inspection, supervision, and administrative management, which shall be implemented on a regular basis. The time, procedure, results, and penalties of on-site inspections will be disclosed on the Permitting Management Information System. The authority will visit the polluter more frequently in case of any prior history of violations.

The original emission permit of the polluter shall be presented in the office or place of production and operation, and shall not be leased, lent, resold, or illegally transferred in any ways. Polluters shall promptly and truthfully release actual emission information, such as the operation condition of pollution treatment facilities and discharging details. For any polluters discharging key pollutants, the annual implementation report shall be compiled, on their own or by an entrusted third party, according to the emission permit, and submitted to the competent authority for environmental protection, in addition to being made available to the public. The competent authority for environmental protection shall supervise the implementation of emission permits and document relevant information so as to establish archive and record management for emission permits. In addition, the information concerning the supervision on the issuance and management of emission permit shall be made available to the public by the competent authority for environmental protection, in accordance with national and local requirements of information disclosure.

Any emission permit acquired through counterfeit information or illegal means shall be revoked. For any actual emission information violating the emission conditions prescribed in the emission permit, the emission permit shall be revoked. If the polluter was forcibly shut down in accordance with relevant laws and regulations, the emission permit shall certainly be revoked. The act of discharging pollutants without an emission permit shall be penalized according to Article 63 of the EP Law, as the executive officer or whoever is directly liable shall be detained for no less than 10 days but no more than 15 days. In case of minor offence, the period of detention shall be no less than five days but no more than 10 days.

Some emission behaviors of the polluter shall be considered as being in violation of the emission permit, and the polluter shall be ordered to stop and correct them within a time limit prescribed by the competent authority for environmental protection, as shown in Box 9.1. Failure to make corresponding corrections shall be punished in accordance with relevant laws and regulations.

Box 9.1 Emission behaviors of the polluter resulting in ceasing operation
- *altering, forging, leasing, lending, reselling or illegally transferring the emission permit in any ways.*
- *failing to fulfill the mandatory setup of the pollutant outlet in accordance with related regulations.*
- *removing or idling pollution prevention and control facilities without the approval of the competent authority for environmental protection.*
- *failing to carry out self-monitoring and disclose environmental information in accordance with related regulations for polluter discharging key pollutants.*

If the emission concentrations exceeds the emission standards specified in the emission permit or the emission amount exceeds the emission cap stipulated in the emission permit, the emissions charges shall be doubled in accordance with the relevant emissions charges standards instituted at the provincial level; if the emission standards and the emission cap were both violated, the emissions charges shall be tripled. For key pollutant(s), if the emission concentration exceeds the emission standards specified in the emission permit or the emission amount exceeds the emission cap stipulated in the emission permit, the polluter shall be ordered to cease operation, in case of serious violation, shut down by the empowered local government.

In case of any of the following emission behaviors, the polluter will not only lose its emission permit, but also be forbidden to re-apply for an emission permit for a year, as shown in Box 9.2.

> **Box 9.2 Emission behaviors of the polluter resulting in canceling emission permit**
> - *discharging pollutants through underground pipes, seepage wells or seepage pits, illegally.*
> - *and tampering or falsifying monitoring data.*
> - *operating pollution prevention and treatment facilities improperly.*
> - *illegally discharging heavy metals and persistent organic pollutants (which will cause serious damage to the environment and human health) exceeding the emission standards for more than 3 times.*

9.4 Pilot Program of Emissions Trading System

In China, the Emissions Trading System has been implemented experimentally since the 1990s, driven by both the central and local governments (Chang & Wang, 2010; Guo, 2018). The launch of the Emission Cap System could be seen as the starting point of the Emissions Trading System, in which the responsibility for the total amount of pollutants was allocated to local governments at various levels through the Target Responsibility System. Under the Emission Cap System, a series of documents were issued by the central government to encourage the practice of emissions trading at local level. In 1994, six cities, including Baotou, Kaiyuan, Liuzhou, Taiyuan, Pingdingshan and Guiyang, were designated as the first batch of cities to undertake the pilot program of emissions trading for air pollutants. In October 2002, the *Pollution Prevention and Control Plan of the Two-Control-Zone Acid Rain Control Zone and Sulfur Dioxide Control Zone during the 10th FYP* was approved by the State Council to propose a pilot project for SO_2 emission trading scheme in the Two-Control-Zone, the Acid Rain Control Zone, and the Sulfur Dioxide Control Zone, respectively. In December 2005, *the Decision on Implementing the Scientific Outlook on Development and Strengthening Environmental Protection* was enacted by the State Council, ensuring that the emission trading of SO_2 emission permit is executed in these areas and industrial sectors fulfill the necessary requirements.

From 2007 to 2013, several areas, including Jiangsu, Zhejiang, Hunan, Hubei, Henan, Hebei, Shanxi, Shaanxi, Inner Mongolia, Tianjin, and Chongqing, were gradually approved by the MOF, MEP, and NDRC to commence the pilot program of emissions trading. Based on the experience accrued from these pilot programs, the *Guiding Opinions Concerning further Promoting the Pilot Program of the Compensable Utilizations of the Emission Permit and Emissions Trading* (the Guiding Opinions 2014) was publicized by the General Office of the State Council in August 2014 to advance the practice of emissions trading. In the 13th FYP (2016–2020), it was further explicitly stated that a sound system be established for compensable utilization of the emission permit and emissions trading. It was also specified in the

Air Pollution Law (2018) that the State shall gradually carry out emission trading of key air pollutants. At the local level, many local governments issued specific regulations to safeguard the emission trading scheme in their jurisdictions. By 2018, around twenty eight provinces and municipalities had implemented their own experimental emission trading programs through the establishment of local regulations and trading platforms (Guo, 2018).

The Guiding Opinions (2014) and several supplementary regulations (*the Interim Measures of Management for the Income from Transferring the Emission Permit* and *the Implementation Measures of Controlling the Emission Permit System to Control Pollutant Emission*) have provided the policy framework for the Emissions Trading System in China. Firstly, under the Emission Permit System, emission permits are allocated to polluters, thereby ensuring that emission trading is arranged between polluters based on the principle of fairness, aiding environmental quality improvement, and optimizing environmental resource allocation. Trading price shall be determined by the trading parties. Some restrictions are imposed on the emission trading:

① the trading shall, in principle, be conducted within the territory of each pilot province/municipality;
② the trading related to water pollutants shall only be conducted within the same drainage basin;
③ thermal power plants, in principle, shall not trade with polluters from other industrial sectors for emission permits of air pollutants;
④ polluters from industrial sectors shall not trade with polluters from agricultural sectors;
⑤ the trading shall not increase the total amount of emissions in the region if such region fails to attain the environmental quality standards set by State (Guo, 2018).

The Chinese government is eager to promote energy saving and pollution reduction by use of a market mechanism. Based on the experience gained from the pilot programs of emissions trading at the local level, the government is considering the development of a cross-regional emissions trading market.

References

Chang, Y.-C., & Wang, N. (2010). Environmental regulations and emissions trading in China. *Energy Policy, 38*(7), 3356–3364. https://doi.org/10.1016/j.enpol.2010.02.006. [2019-09-05].

Guo, H. (2018). China's experiment of emission permits trading. *Environmental Development, 26*, 112–122. https://doi.org/10.1016/j.envdev.2018.02.001. [2019-09-05].

Chapter 10
Emissions Cap System

The Emissions Cap System is one of the direct or command-and-control regulations, in which the overall emissions limits are set up by governments, and then allocated to different areas or sectors, as allowances, so as to fulfill pollution control. Through the gradual expiration of a certain percentage of existing allowances, the limits shall be reduced constantly.

10.1 Establishment and Development of the Emissions Cap System

Prior to the introduction of the Emissions cap System, there were two major measures of environmental governance in China, a series of emission standards and potential penalties and fines for noncompliance. During the 6th FYP (1981–1985), several research programs on water environmental capacity were implemented in various rivers, including the Yangtze River, the Huangpu River (Shanghai), and the Xiang River (Hunan Province). During the 7th FYP (1986–1990), large-scale research programs on water environmental capacity, water pollution control planning, lake eutrophication prevention and control, and emissions cap were deployed in six major watersheds (including the Yangtze River, Pearl River, and Huai River) and 25 lakes (such as the Dian Lake, Chao Lake, Tai Lake, etc.). Through these research programs, various water quality models were established for different water bodies to analyze and calculate the water environmental capacity so as to determine the emissions cap of key pollutants, and to institute a comprehensive governance plan for the water environment.

In April 1989, the approach of implementing a concentration control and emissions cap concurrently was proposed by the NEPA during the Third National Conference on Environmental Protection, as well as the strategy of transiting from concentration control to emissions cap. This was the first time that the concept of emissions cap was proposed at the national policy level.

In July 1996, during the Fourth National Conference on Environmental Protection, it was determined that the emissions cap system shall be implemented throughout the country during the 9th FYP (1996–2000). In August of the same year, the State Council issued the *Decision on Several Issues Concerning Environmental Protection*, which required that: ① all industrial sources shall be in compliance with national or local emission standards by 2000; ② the total amount of major pollutants shall be limited to the allowance predetermined by the central government for every province, autonomous region, and municipality directly under the central government, and the trend of environmental pollution and ecological degradation shall be basically under control; ③ the air and surface water quality in major cities (including municipalities directly under the central government, provincial capitals, cities with special economic zones, open cities in coastal zones, and key tourist cities) shall be in compliance with relevant standards stipulated by the central government, in accordance with their functional zones. Further, in September of the same year, *the Ninth Five-Year Plan for the National Economic and Social Development of the People's Republic of China and the Long-Term Goals Objectives for 2010*, jointly formulated by the NEPA, SPC, and SETC, was approved by the State Council. As explicitly required by Annex 1, the *Plan for Emissions Cap of Major Pollutants Nationwide During the 9th FYP*, the level of 12 indicators (including soot, industrial dust, sulfur dioxide, chemical oxygen demand, petroleum pollutants, cyanide, arsenic, mercury, lead, cadmium, hexavalent chromium, and industrial solid waste) shall be reduced by 10%–15% by 2010, as show in Table 10.1. In addition, the following objectives were

Table 10.1 The emissions targets of 12 indicators during the 9th FYP

Indicators	Emissions in 1995	Targets for 2000	Actual emissions in 2000
Soot/10^4 ton	1744	1750	1165
Industrial dust/10^4 ton	1731	1700	1092
Sulfur dioxide, SO_2/10^4 ton	2370	2460	1995
Chemical oxygen demand, COD/10^4 ton	2233	2200	1445
Petroleum pollutants/ton	84,370	83,100	34,000
Cyanide/ton	3495	3263	923.8
Arsenic/ton	1446	1376	578.7
Mercury/ton	27	26	10.1
Lead/ton	1700	1670	655.2
Cadmium/ton	285	270	138.5
Hexavalent chromium, Cr^{6+}/ton	670	618	119.7
Industrial solid waste/ton	6170	5995	3186.1

Data Source China Annual Report of Environmental Statistics (2000).

10.1 Establishment and Development of the Emissions Cap System

Table 10.2 The emissions targets of 6 indicators during the 10th FYP

Indicators	Emissions in 2000	Targets for 2005	Actual emissions in 2005
Soot/10^4 ton	1165	1100	1182.5
Industrial dust/10^4 ton	1092	900	911.2
Sulfur dioxide, SO_2/10^4 ton	1995	1800	2549.3
Chemical oxygen demand, COD/10^4 ton	1445	1300	1414.2
NH_3–N/10^4 ton	–	165	149.8
Industrial solid waste/ton	3186.1	2900	1654.7

Data Source National Bureau of Statistics.

also proposed: implementing the emissions cap throughout the country, achieving the attainment of all functional zones in 47 major cities, and ensuring that all industrial emissions shall be in compliance with relevant emission standards (the so-called, One-Control-and-Two-Attainment), by 2000.

As explicitly specified in the *National Environmental Protection Plan during the 10th FYP*, promulgated by the SPEA, SPC, SETC, and MOF in December 2011, the number of regulated indicators was reduced from 12 to 6, including soot, industrial dust, sulfur dioxide, chemical oxygen demand, ammonia nitrogen, and industrial solid waste (as show in Table 10.2); the levels of these indicators shall be decreased by 10% in comparison to those in 2000; the levels of other indicators in industrial wastewater (such as heavy metals, cyanide, petroleum, and other pollutants) shall be effectively contained; and the amount of SO_2 emissions in the Two-Control-Zone (Acid Rain Control Zones and Sulfur Dioxide Pollution Control Zones, see Table 6.1) shall be reduced by 20% in comparison to that in 2000, while the acidity of precipitation and the frequency of acid rain shall be reduced. In addition, the *Allocation Plan for the Emissions Cap of Major Pollutants during the 10th FYP* was instituted and implemented.

In March 2003, the *Notice on Several Issues Concerning the Allowance Allocation of the Emissions Cap of Major Pollutants from Construction Projects* was enacted by the SEPA to ensure that in the Project EIA, the pattern and amount of major pollutants from construction projects, as well as the impact of any construction project once it is completed and normally operated, shall be carefully evaluated and predicted, and the countermeasures of pollution prevention and control and schemes of emissions reduction shall be proposed, which would serve as the base to determine the allowance allocation of the emissions cap of major pollutants for the project. In order to ensure the effective implementation of the Emissions Cap System, the *Notice on the Scheme of Determining the Environmental Capacity of Surface Water and Ambient Air Nationwide* was published by the SEPA in August of the same year to regulate that the determination of environmental capacity nationwide shall be

Table 10.3 Emissions targets of 2 binding indicators during the 11th FYP

Indicators	Emissions in 2005	Targets for 2010	Actual emissions in 2010
$SO_2/10^4$ ton	2549.3	2295	2185.1
$COD/10^4$ ton	1414.2	1270	1238.1

Data Source China Annual Report of Environmental Statistics (2010).

based upon the scientific understanding and accurate estimation of the environmental capacity in different regions, watersheds, and cities.

During the 11th FYP period (2006–2010), environmental protection planning in China experienced a transition from soft constraints to hard constraints, in which it was sought to reduce the emissions of the two binding indicators, SO_2 and COD, by 10% by 2010 based on the levels in 2005, as shown in Table 10.3. In 2007, various measures of statistics, monitoring, and evaluation on emissions cap were enacted by the State Council, as shown in Box 10.1.

Box 10.1 Various measures of statistics, monitoring and evaluation on emissions cap

- the *Measures of Monitoring the Emissions Reduction of Major Pollutants* (the State Council [2007] No. 36).
- the *Measures of Evaluating the Emissions Reduction of Major Pollutants during the 11th FYP (on Trial)* (the SEPA [2007] No. 124).
- the *Detailed Rules of Auditing the Emissions Reduction of Major Pollutants (on Trial)* (the SEPA [2007] No. 183).
- the *Notice Concerning the Implementation of Quarterly Reporting System for Emissions Reduction of Major Pollutants* (the SEPA [2007] No. 131).

During the 12th FYP (2011–2015), along with SO_2 and COD, ammonia nitrogen (NH_3–N) and nitrogen oxides (NO_x) were added as binding indicators, as shown in Table 10.4. Agricultural sources and motor vehicles were included in emission control. In addition, in the special plans for key areas, watersheds, and marine areas confirmed by the State, emissions of key heavy metals, total nitrogen, total phosphorus, and other pollutants were to be better controlled.

Table 10.4 Emissions targets of 4 binding indicators during the 12th FYP

Indicators	Emissions in 2010	Targets for 2015	Actual emissions in 2015
$SO_2/10^4$ ton	2185.1	2086.4	1859.1
$NO_x/10^4$ ton	1852.4	2046.2	1851.0
$COD/10^4$ ton	1238.1	2347.6	2223.5
$NH_3-N/10^4$ ton	120.3	238.0	229.9

Data Source China Annual Report of Environmental Statistics (2015).

10.2 Objective Classifications of Emissions Cap

Emissions cap is the amount of pollutants allowed to be discharged within a specific area during a certain time period; it is determined according to regional environmental capacity, environmental target, previous emissions history, and cost benefit analysis.

10.2.1 Emissions Cap Based on Environmental Capacity

Environmental capacity is defined as a property of the environment and its ability to accommodate a particular activity or rate of an activity without any unacceptable impact (GESAMP, 1986). Based on environmental capacity, the emissions cap, an environmental threshold, is referred to as the amount of pollutants that the environment can accommodate without impairing its functions. Once the amount of pollutants exceeds the environmental threshold, the functions of the environment are impaired. However, due to the complex process of migration and transformation of pollutants in the environment, it is very difficult to calculate the emissions cap of pollutants based on environmental capacity (Hu et al., 2018).

10.2.2 Emissions Cap Based on Environmental Target

Owing to the difficulty in implementing the emissions cap based on environmental capacity, currently, the most common method is to adopt the environmental target or corresponding standards as the basis for determining environmental capacity. Therefore, the emissions cap in a region shall be limited to the maximum amount of pollutants so as to ensure that the environmental quality is in compliance with relevant standards. Generally, simulation models shall be applied to analyze the contribution to environmental impacts from the original amount of pollutants, and the impact from additional amounts of pollutants, which is basically identical to the existing process of environmental impact assessment.

10.2.3 *Emissions Cap Based on Previous Emissions*

This is the method presently adopted in China. For example, as stipulated in the 10th FYP (2001–2005), the emissions cap for 2005 shall be reduced by 10% from the level of 2000. Further, as specified in the 11th FYP period (2006–2010), the emissions cap for 2010 shall be reduced by 10%, based on the level of 2005. Therefore, it is essential to have a complete and accurate historical emissions inventory (Ge, Chen, Wang, & Long, 2009).

10.2.4 *Emissions Cap Based on Cost Benefit Analysis*

From the perspectives of economists, the most economically efficient level for the emissions cap is where marginal abatement costs are equal to marginal benefits from the emissions reduction. However, this level is often difficult to determine due to uncertain information. More generally, the emissions cap should be set at a level that is expected to address environmental and health concerns at an acceptable cost (EPA, 2003).

10.3 Pollutants Classification of Emissions Cap

As per the provisions contained in Article 44 of the EP Law (2014), the emissions cap system for major pollutants shall be implemented throughout the country. The allowance of major pollutants shall be determined by the State Council and allocated by local governments at the provincial level. In addition to complying with national and regional emission standards, polluters shall abide by the allowance of major pollutants that have been allocated to them. In case of noncompliance with the allocated allowance of major pollutants or nonattainment of environmental quality targets predetermined by the State, the competent authority for environmental protection at the provincial level or above shall suspend the review and approval process for the EIA documents of the construction project with a new increment of major pollutants within those areas.

10.3.1 Emissions Cap of Water Pollutants

As stipulated in Article 18 of the Water Pollution Prevention Law (2008), the emissions cap system for major water pollutants shall be implemented. For local governments at the provincial level, the emissions of major water pollutants within the administrative districts shall be diminished according to the regulations issued by the State Council, and the allowance of major water pollutants designed by the State shall be concretely allocated to subordinate cities and counties. For city and county governments, the allowance of major water pollutants distributed by provincial government shall be rationally allotted to pollution units. Based on the status quo of the water quality and the needs of water pollution prevention and control within the administrative districts, local governments at the provincial level shall determine the major water pollutants and local emissions cap on condition that the requirements set by the State shall be fulfilled, first. In case the assigned allowance is exceeded, the competent authority for environmental protection at the local government level shall suspend the review and approval process for the EIA documents of the construction project with a new increment of major water pollutants within those areas.

10.3.2 Emissions Cap of Air Pollutants

As specified in Article 21 of the Air Pollution Prevention Law (2015), the emissions cap system for major air pollutants shall be implemented. After soliciting opinions from local governments at the provincial level and other relevant departments of the State Council, the competent authority for environmental protection of the State Council, in conjunction with the competent authority for comprehensive economic management of the State Council, shall submit the emissions cap for major air pollutants to the State Council for approval. Subsequently, local governments at the provincial level shall control and reduce the emissions of major air pollutants according to the allowance assigned by the State Council. Further, the emissions trading system for major air pollutants shall be gradually implemented throughout the country. In case the allocated allowance of major pollutants cannot be accomplished or the air quality target preset by the State cannot be achieved, the competent authority for environmental protection at the provincial level shall suspend the review and approval process for the EIA documents of the construction project with a new increment of major pollutants within those areas, in addition to an interview with the chief officer of the regional government. The results of the interview shall be disclosed to the general public.

10.3.3 Emissions Cap of Pollutants Discharged into the Sea

As stated in Article 10 of the Marine Protection Law (2013), the emissions cap system shall be established and implemented in key marine areas to regulate the amount of major pollutants to be discharged into the sea and to allocate the allowance to major pollution sources. Specific measures shall be formulated by the State Council.

10.4 Measures of Emissions Cap

The design and implementation of the Emissions Cap System, which can be quite complicated, can be divided into several stages, including schematic design, target determination, plan institution, and practical implementation. For example, the compiling process of the plan for the emissions cap of major pollutants during the 12th FYP is illustrated in Fig. 10.1.

10.4.1 Evaluations on the Emissions Cap System During the 11th FYP

Analyzing and assessing the implementation of the emissions cap of major pollutants during the 11th FYP was of significant importance for the purpose of instituting the criteria during the 12th FYP, based on the actual emissions in 2010. The main contents of the analysis and assessment are listed in Box 10.2.

> **Box 10.2 Main content of analysis and assessment on the emissions cap system during the 11th FYP**
> - *the accomplishment of national and regional targets of the emissions cap during the 11th FYP.*
> - *the implementation of major measures of emissions reduction.*
> - *the institution and implementation of supplement measures of emissions reduction.*
> - *existing practical problems and recommendations.*

10.4 Measures of Emissions Cap

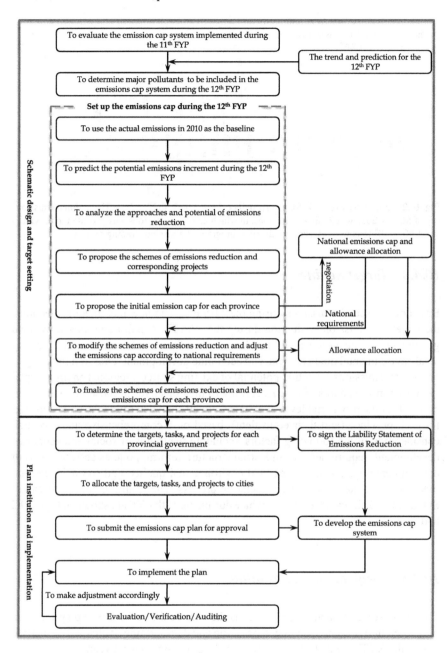

Fig. 10.1 Compiling process of the plan for the emissions cap of major pollutants during the 12th FYP

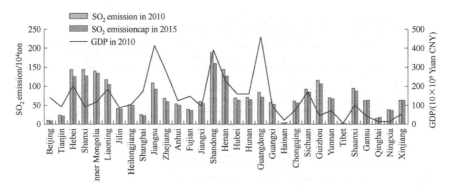

Fig. 10.2 Emission cap of SO_2 for each province during the 12th FYP *Note* The national emissions cap of SO_2 by 2015 was 2.0864×10^7 tons, in which 2.0674×10^7 tons was allocated to provinces and 1.90×10^5 tons was reserved for the pilot program of emissions trading of SO_2

10.4.2 Target Setting

Based on the actual emissions during the 11th FYP and new increment projections, the targets of emissions reduction during the 12th FYP shall be proposed by local governments at the provincial level, taking various factors into consideration, such as local environmental quality, social and economic development, emission intensity of pollutants, emissions reduction potential of existing sources, regional environmental carrying capacity, and so forth. According to the objectives proposed by local governments at the provincial level, the State Council shall determine the emissions cap for each province (and their equivalent), based on a comprehensive and thorough consideration of the following factors, including the trends of social and economic development, requirements of industrial structure reform, policies and standards of environmental management, regional environmental quality, and so on. Therefore, this could be quite a repetitive negotiation process. The resulting emissions cap of SO_2 and COD allocated to each province during the 12th FYP are shown in Figs. 10.2 and 10.3.

10.4.3 Verification and Auditing

The verification and auditing process, as the core of the emissions cap system, can be divided into two parts, on-site verification and data auditing, which will be carried out by the MEE and its Centers of Regional Inspection, twice a year.

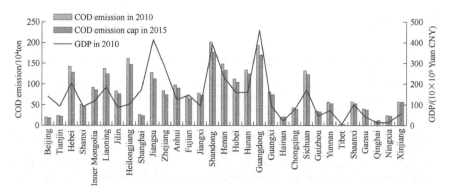

Fig. 10.3 Emission cap of COD for each province during the 12th FYP *Note* The national emissions cap of COD by 2015 was 2.3476×10^7 tons (including 1.2219×10^7 tons for industrial and domestic sources), in which 2.3352×10^7 tons was allocated to provinces (including 1.2146×10^7 tons for industrial and domestic sources) and 1.24×10^5 tons was reserved for the pilot program of emissions trading of COD

10.5 Final Remarks

The legal status of the emissions cap system was established by the enactment of the EP Law (2014) enacted during the 12th FYP. Due to a sturdy evaluation mechanism, the emissions cap system has at present become a strongly enforced environmental protection scheme in China. Meanwhile, the deployment of the target responsibility system has effectively promoted the concrete implementation of the emissions cap system.

Within the emissions cap system, the local governments at all levels shall be subject to the liability of the fulfillment of emissions reduction, which shall be directly applied to the performance evaluation of the governments, and the appointment, promotion, and dismissal of government officials. The central government will evaluate the performance of provincial governments according to the attainment of their annual targets. In case of any unmet annual targets, regional restrictions on new projects shall be applied, local governments will be required to make corresponding improvements within certain time limits, and the government officials shall be interviewed for corresponding responsibility. Likewise, based on emissions data, provincial governments shall conduct performance evaluation on city and county governments so as to promote the development of both economic growth and environmental protection in harmony.

References

EPA. (2003). Tools of the trade: A guide to designing and operating a cap and trade program for pollution control.

Ge, C., Chen, J., Wang, J., & Long, F. (2009). China's total emission control policy: A critical review. *Chinese Journal of Population Resources and Environment, 7*(2), 50–58. https://doi.org/10.1080/10042857.2009.10684924. [2019-09-11].

GESAMP. (1986). Environmental capacity: An approach to marine pollution prevention: IMO/FAO/UNESCO/WMO/WHO/IAEA/UNEP Joint Group of Experts on the Scientific Aspects of Marine Pollution. GESAMP Reports and Studies.

Hu, Q., Li, X., Lin, A., Qi, W., Li, X., & Yang, X. J. (2018). Total emission control policy in China. *Environmental Development, 25,* 126–129. https://doi.org/10.1016/j.envdev.2017.11.002. [2019-09-12].

Chapter 11
Joint Pollution Prevention and Control

As explicitly expressed in Article 20 of the EP Law (2014), the mechanism of joint pollution prevention and control shall be established for cross-regional environmental pollution and ecological degradation in key regions and watersheds, where the unified planning, standards, monitoring, and measures of pollution prevention and control shall be applied. Joint pollution prevention and control is an effective scheme to greatly improve regional environmental quality through cross-regional cooperation and collaboration (Su & Yu, 2019). Taking air pollution as an example, the working mechanism of joint pollution prevention and control shall be illustrated in the sections.

11.1 The Need for Joint Air Pollution Prevention and Control

At present, the characteristics of air pollution in China have been undergoing significant changes, reflecting the prominent features of complication and regionalization, especially in highly industrialized and urbanized regions, and contiguous agglomerated urban areas. Due to huge resource demands and high energy consumption in urban agglomerations, along with the high intensity and concentration of atmospheric pollutants discharged, the ambient air quality in urban areas is seriously degraded, more frequently and intensively, owing to heavy air pollution. In addition, the transportation, conversion, and coupling of primary air pollutants between neighboring cities results in the constant increase of fine particulate matter (PM) concentration, ozone (O_3) concentration, the frequency of acid rain, and the occurrence of haze (Wu, Xu, & Zhang, 2015; Zhang, Ma, Qin, & Li, 2018). These problems are too tough and complicated to be solved effectively by solely relying on individual cities, alone. Therefore, there is an urgent need to surpass the limits of administrative boundaries by adopting the measures of joint air pollution prevention and control, in which all resources are utilized comprehensively and the liability shall be strictly allocated and implemented so as to form a joint force of pollution prevention and control (Feng & Liao, 2016; Guo & Lu, 2019).

In May 2010, the *Guiding Opinions Concerning the Promotion of Joint Regional Air Pollution Prevention and Control Work for Improving Regional Air Quality*, compiled by the MEP, were enacted by the General Office of the State Council to facilitate the implementation of joint air pollution prevention and control. In December 2012, the *Air Pollution Prevention and Control for Key Regions during the 12th FYP* was jointly instituted by the MEP, NDRC, and MOF to propose several key tasks of air pollution prevention and control during the 12th FYP, as listed in Box 11.1. In September 2013, the overall objectives of the coordination between regional collaboration and administrative management and the implementation of regionalized governance in different timeframes were clearly defined in the Action Plan for Air Pollution, in addition to other goals, such as a general improvement in the national ambient air quality, a significant reduction in the number of days with heavy pollution, and a considerable improvement in the regional ambient air quality in the Beijing-Tianjin-Hebei area. As specified in Article 2 of the Air Pollution Law, in order to prevent and control air pollution, it is necessary to strengthen the comprehensive prevention and control of air pollution from coal burning, industries, motor vehicles and vessels, street dust, agriculture, and so on, to implement the joint prevention and control of regional air pollution, and to carry out the collaborative control of PM, SO_2, NO_x, volatile organic compounds (VOCs), NH_3–N, and so forth.

Box 11.1 Key tasks specified in the Air Pollution Prevention and Control for Key Regions during the 12th FYP
- *To plan comprehensively for regional environmental resources.*
- *To optimize the industrial structure and layout.*
- *To phaseout outdated production.*
- *To improve the utilizations of clean energy.*
- *To control the regional total coal consumption.*
- *To enhance air pollution governance.*
- *To implement collaborative control of various pollutants.*
- *To innovate regional management mechanism.*
- *To elevate the management capacity of joint pollution prevention and control.*
- *To highlight the measures of joint prevention and control of regional pollution.*

11.2 Key Regions and Key Points of Joint Air Pollution Prevention and Control

Several key regions have been earmarked for the deployment of joint air pollution prevention and control, including the Beijing-Tianjin-Hebei Region, the Yangtze River Delta Region, and the Pearl River Delta Region. In addition, the joint air pollution prevention and control shall be proactively promoted in the following regions, such as the central part of Liaoning province, the Shandong Peninsula, Wuhan and its surrounding areas, the Changsha-Zhuzhou-Xiangtan Region, the Chengdu-Chongqing Region, and the west coast of the Taiwan Straits. Nevertheless, local governments in other regions are expected to promote the joint air pollution prevention and control based on the actual local circumstances.

Within the joint air pollution prevention and control, there are "four keys," including the key pollutants (SO_2, NO_x, PM, VOCs, and so on), key industries (thermal power, steel, nonferrous metals, petrochemicals, cement, and chemical industries), key enterprises (those that have a greater impact on regional air quality), and key problems (acid rain, haze, photochemical smog, etc.) (Wang & Zhao, 2018).

11.3 Key Tasks of Joint Air Pollution Prevention and Control

Several key tasks of Joint Air Pollution Prevention and Control were proposed, as illustrated below.

11.3.1 Optimizing the Regional Industrial Structure and Layout

In order to optimize the regional industrial structure and layout, the application of stricter environmental standards to industries, optimization of the regional industrial layout, and promotion of technological advancements and structural adjustments is of paramount importance.

(1) Application of stricter environmental standards for key industries in key regions

For key industries in key regions, special emissions limits on air pollutants shall be applied to strictly regulate any new construction and expansion of thermal power plants in key regions, except for large-scale thermal power plants and combined heat and power (CHP) plants. In urban areas of prefecture-level cities, any new construction and expansion of thermal power plants is forbidden, with the exception of CHP plants. In addition, in key regions, the regional consultation meeting shall be held for Project EIA of the construction projects of key industries. Moreover, it is of great importance to strengthen PEIA (EIA for Plan) of the regional industrial development plan so as to stringently control the expansion projects of certain industries with excess capacity, including the iron and steel industry, cement industry, flat glass industry, traditional coal chemical industry, polysilicon industry, electrolytic aluminum industry, shipbuilding industry, and so forth.

(2) To optimize the regional industrial layout

In order to prevent and control pollution transfer during industrial transfer, it is essential to establish the environmental monitoring and regulatory mechanisms to supervise the transferring process. In urban and suburban areas, new construction and expansion of the following industries shall be prohibited, including iron and steel industry, nonferrous metals industry, petrochemical industry, cement industry, chemical industry, and other enterprises with heavy pollution. Furthermore, in urban areas, relocation and transformation shall be applied to existing enterprises with heavy pollution through industrial restructuring.

(3) To promote technological advancements and structural adjustment

Various tasks shall be performed, for example, to improve the standards and evaluation indicators of cleaner production for key industries so as to enhance the audit and assessment of cleaner production for key enterprises, to greatly promote the technologies of cleaner production so as to encourage enterprises to adopt advanced technologies of cleaner production, to speed up the industrial restructuring so as to phase out outdated enterprises within key industries, including electricity, coal, steel, cement, nonferrous metals, coke, paper, tan, and printing and dyeing.

11.3.2 *Intensifying the Extent of Pollution Prevention and Control on Key Pollutants*

(1) To strengthen the emissions cap system of SO_2

- To elevate the desulfurization efficiency of thermal power plants.
- To improve the franchise system of the desulfurization facilities for thermal power plants.
- To upraise the requirement of SO_2 emissions reduction for the iron and steel industry, petrochemical industry, nonferrous metals industry, and so forth.
- To boost the desulfurization of industrial boilers.
- To perfect the emissions charges system of SO_2.
- To institute the target of regional SO_2 emissions reduction.

(2) To enhance NO_x emissions reduction

- To establish the emissions cap system of NO_x.

- To install a flue gas denitrification facility while newly constructing, expanding, and reconstructing the thermal power plants, according to the emissions standards and requirements of the EIA documents of the construction project.
- To require all thermal power plants in key regions to be equipped with a flue gas denitrification facility during the 12th FYP.
- To reserve enough space for a flue gas denitrification facility for thermal power plants in other areas.
- To promote low NO_x-emissions combustion technology for industrial boilers.
- To deploy NO_x pollution prevention and control on the iron and steel industry, petrochemical industry, and chemical industry.

(3) To intensify PM pollution prevention and control

- To apply bag house and other efficient dedusting techniques to cement plants, thermal power plants, and any enterprises with industrial boilers.
- To enhance the environmental management of construction sites.
- To prohibit the use of bulk cement, and on-site mixing of concrete and mortar.
- To adopt fencing, covers, and other dust control measures at construction sites.
- To improve the road cleaning so as to improve urban road cleanliness.
- To implement the "no bare ground in open air" scheme so as to reduce the area of barren land in urban regions.

(4) To employ VOCs pollution prevention and control

- To implement pollution governance on VOC emissions from the production process of paint spraying, petrochemical, shoemaking, printing, electronics, and dry cleaning, according to the relevant technical specifications.
- To promote oil and gas pollution governance at gas stations.
- To complete the modification of the oil and gas recovery system for existing oil depots, gas stations, and oil tankers in the key regions, on schedule, and to ensure the normal operation of all facilities.
- To install the oil and gas recovery system for new oil depots, gas stations, and oil tankers prior to being put into use.
- To strictly control the obnoxious gases emissions from the catering industry in urban areas.

11.3.3 Promoting the Comprehensive Utilization of Clean Energy

(1) To stringently control pollution emissions from coal-burning

- To strictly control the construction of coal-burning projects in key regions.
- To employ the pilot program of regional total coal consumption control.

- To facilitate the construction of distribution centers for low-sulfur and low-ash coal.
- To elevate the proportion of coal washing and screening.
- To prohibit direct coal-burning with sulfur content higher than 0.5% for enterprises without desulphurization facilities in key regions.
- To designate zones where the burning of heavy pollution fuels is forbidden and to gradually expand these zones.
- To prohibit scatter burning of raw coal.
- To construct projects that demonstrate multiple control technologies of desulfurization, denitrification, dedusting, and mercury removal for flue gas of thermal power plants.

(2) To robustly promote clean energy

- To improve the urban energy structure by promoting the utilization of clean energy (natural gas, liquefied petroleum gas, coal gas, and solar energy) so as to increase the proportion of clean energy in urban areas.
- To boost the action of clean energy and encourage the demonstration of clean energy application.
- To promote energy conservation in industries, transportation, and construction and to increase energy efficiency.
- To speed up the development of rural clean energy by encouraging the comprehensive utilization of crop straw, promoting biomass fuel technologies, and developing rural biogas.
- To prohibit the outdoor burning of crop straw and other agricultural waste to ensure clean ambient air quality in the locations around cities, traffic arteries, and airports.
- To encourage the applications of energy-saving stoves so as to gradually phase out traditional high-pollution stoves areas.

(3) To actively develop centralized heating in urban areas

- To promote the construction of centralized heating in urban areas, and to strengthen the grid-connection of urban heating boilers so as to continuously expand the coverage of central heating in urban areas.
- To enhance the desulfurization, denitrification, and efficient dedusting of flue gas from centralized heating boilers.
- To develop clean coal technologies, and to promote the demonstration of clean coal centralized heating boilers.
- To forbid the construction of small low-efficiency, high-polluting coal-burning boilers in urban and suburban areas, and to gradually dismantle existing small coal-burning boilers.

11.3.4 Intensifying Pollution Prevention and Control of Motor Vehicles

(1) To reduce the emissions from motor vehicles

- To strictly implement national emissions standards for motor vehicles.
- To improve the approval system of environmentally friendly motor vehicles.
- To prohibit the production, sales, and registration of motor vehicles that are not in compliance with the national emissions standards for motor vehicles.
- To promote trade-ins so as to speed up the process of phasing out yellow-tag motor vehicles and low-speed trucks.
- To develop new energy vehicles.

(2) To perfect the environmental management system of motor vehicles

- To strengthen periodic inspections of exhaust emissions of motor vehicles.
- To implement environmental label management for motor vehicles.
- To deploy special measures in the case of motor vehicles that are not in compliance with the national emissions standards for motor vehicles.
- To enhance supervision and management of organizations to perform periodic inspections on exhaust emissions of motor vehicles so as to promote healthy development.
- To intensify the capacity building of periodic inspection of motor vehicles.
- To establish an environmental management information system for motor vehicles.
- To deploy research on tax policies to facilitate pollution prevention and control of motor vehicles.

(3) To speed up clean fuel processes for vehicles

- To promote low-sulfur fuels for motor vehicles and to speed up the process of modifying oil refinery facilities so as to increase the market supply of higher quality motor vehicles fuels.
- To institute and implement the fourth and fifth stages of the national motor vehicles fuels standards and the limits on the harmful substances content of motor vehicles fuels.
- To strengthen the management of motor vehicles fuels detergents.

(4) To vigorously develop public transportation

- To enhance the infrastructure of urban transportation.
- To deploy the strategy of public transportation first.
- To speed up the construction of dedicated lanes for buses and trolley buses and to establish the signal system of public transportation first.
- To improve the ease of walking and cycling for the public so as to encourage the public to choose green travel method.

11.3.5 Improving the Regional Air Quality Monitoring System

(1) To strengthen air quality monitoring in key regions

- To increase the capacity of air quality monitoring.
- To optimize the locations of air quality monitoring stations in key regions.
- To carry out air pollutants monitoring (acid rain, fine PM, and O_3), and air quality monitoring on both sides of urban roads.
- To institute the forecast, early warning and emergency response plans of air pollution accidents.
- To improve the environmental information disclosure system so as to achieve monitoring information sharing in key regions.
- To basically establish an air quality monitoring network in key regions by the end of 2011.

(2) To improve the air quality evaluation index system

- To accelerate the revision of the air quality evaluation index system.
- To enhance the air quality evaluation method of O_3 and fine PM.
- To increase the corresponding evaluation indices.

(3) To strengthen classification management of urban air quality

- To institute air quality compliance schemes for cities that have not achieved compliance with secondary air quality standards so as to ensure the achievement of air quality improvement goals on schedule.
- To report and implement compliance programs in key environmental protection cities.
- To institute air quality improvement schemes for cities that have attained the secondary air quality standards so as to prevent any deterioration in air quality standards.

(4) To enhance the regional supervision and enforcement of environmental laws

- To identify and publicize the list of key enterprises after consultation meetings between the MEP and local relevant departments and agencies.
- To deploy joint law enforcement and inspection of regional air quality.
- To implement centralized pollution governance for enterprises with illegal emissions.
- To strengthen the supervisory monitoring of key enterprises and to promote the installation of online monitoring systems of pollution sources by a local competent authority for environmental protection.
- To require all key enterprises to install online monitoring systems that shall be connected to the networks of the local competent authority for environmental protection by the end of 2012.

References

Feng, L., & Liao, W. (2016). Legislation, plans, and policies for prevention and control of air pollution in China: Achievements, challenges, and improvements. *Journal of Cleaner Production, 112*, 1549–1558. https://doi.org/10.1016/j.jclepro.2015.08.013. [2019-09-19].

Guo, S., & Lu, J. (2019). Jurisdictional air pollution regulation in China: A tragedy of the regulatory anti-commons. *Journal of Cleaner Production, 212*, 1054–1061. https://doi.org/10.1016/j.jclepro.2018.12.068. [2019-09-19].

Su, Y., & Yu, Y.-q. (2019). Spatial association effect of regional pollution control. *Journal of Cleaner Production, 213*, 540–552. https://doi.org/10.1016/j.jclepro.2018.12.121. [2019-09-19].

Wang, H., & Zhao, L. (2018). A joint prevention and control mechanism for air pollution in the Beijing-Tianjin-Hebei region in china based on long-term and massive data mining of pollutant concentration. *Atmospheric Environment, 174*, 25–42. https://doi.org/10.1016/j.atmosenv.2017.11.027. [2019-09-19].

Wu, D., Xu, Y., & Zhang, S. (2015). Will joint regional air pollution control be more cost-effective? An empirical study of China's Beijing–Tianjin–Hebei region. *Journal of Environmental Management, 149*, 27–36. https://doi.org/10.1016/j.jenvman.2014.09.032. [2019-09-23].

Zhang, N.-N., Ma, F., Qin, C.-B., & Li, Y.-F. (2018). Spatio temporal trends in PM2.5 levels from 2013 to 2017 and regional demarcations for joint prevention and control of atmospheric pollution in China. *Chemosphere, 210*, 1176–1184. https://doi.org/10.1016/j.chemosphere.2018.07.142. [2019-09-23].

Chapter 12
Public Participation and Environmental Information Disclosure

Public participation in environmental management refers to public participation in environmental decision-making in various forms and through diverse channels, according to law. The purposes of public participation in environmental management are the promoting of the effective implementation of environmental administrative power, increasing the public's environmental behavior and awareness, assisting the administrative authorities to oversee any activities violating the environmental regulations, and supervising the administrative authorities' fulfillment of duties.

As the bridge that connects collective environment rights and individual environment rights, the right to participate involves the establishment of a negotiation mechanism and a consultation mechanism within the system for diverse interest groups through national legislation. Public participation provides various interest groups with the opportunity to express their own interests and demands so as to facilitate the balancing of different interests and to reduce the social contradictions provoked by the conflicts of interest within the environmental protection system. In addition, as the basic requirement of democratic governance and the rule of law, public participation and a supervision mechanism would prevent any environmental damage resulting from the illegal actions and misconduct of administrative authorities.

12.1 Legislative Process of Public Participation in Environmental Management

In China, the legal basis of public participation in environmental management has long been rooted in the Constitution. Since the reform and opening up of the Chinese economy, along with extensive social development and rapid economic growth, all kinds of laws, regulations, provisions, rules and instructions have been promulgated to ensure the rights and obligations of public participation in environmental management.

As stipulated in Article 2 of the Constitution, all the power of the People's Republic of China is empowered by the people. The institutional framework for people to exercise their rights includes the NPC and the local people's congress at all levels. Through various channels and forms, the people manage state affairs as well as the economic, social and cultural activities, in accordance with the laws. People's constitutional right to participate in environmental protection and management was clearly enshrined in this provision. Further, as specified in Article 41 of the Constitution, citizens of the People's Republic of China have the right to level criticism against and propose suggestions to any state agencies or state officials, and to lodge a complaint, make an accusation, or file a lawsuit against any illegal action or misconduct on the part, of any state agencies or state officials.

As explicitly stated in Article 6 of the EP Law (1989), every institution and individual shall have the obligation to protect the environment, and have the right to accuse, or file a lawsuit against, any institutions and individuals for polluting and damaging the environment.

In Chapter 20 of *China's Agenda 21—White Paper on China's Population, Environment and Development in 21st Century*, it was suggested that the realization of sustainable development goals must rely on the support from, and participation of, public and social groups; and the ways and degrees of participation of the public, social groups, and organizations will determine the progress of the realization of sustainable development goals. Social groups and the public must not only be involved in the decision-making process related to the environment and development, especially in those communities where the decisions taken may affect their lives and work, but also be involved in the supervision of the execution of these decisions.

Many detailed requirements were indicated in Article 7 of the *Notice on Strengthening the Management for EIA for Construction Projects Loaned by the International Financial Organizations*, jointly issued by the NEPA, SPC, MOF, and the People's Bank of China, in June 1993, as shown in Box 12.1.

Box 12.1 Detail requirements of public participation specified in the Notice on Strengthening the Management for EIA for Construction Projects Loaned by the International Financial Organizations (the IFO)

- *Public participation is an essential part of environmental impact assessment, and shall be presented in a special chapter of the environmental impacts assessment report.*
- *Public participation can be carried out during compiling and reviewing the outlines of environmental impact assessment or examining the environmental impact assessment report.*
- *According to China's current situation, public participation can be implemented in two ways: ① the opinions and suggestions of the representatives from the People's Congress, the People's Political Consultative Conference, the environmental non-governmental organizations, academic institutes,*

12.1 Legislative Process of Public Participation in Environmental Management

residents committees, and village committees located at the sites of construction projects loaned by the IFO shall be seriously considered by the construction units and the competent authority for environmental protection; ② *the People's Congress, the People's Political Consultative Conference, or the environmental non-governmental organizations shall solicit the opinions and suggestions of the public from the affected areas through various means, including the questionnaire (the Public Opinions Consulting Form), forum, participating in the review meetings of the outlines of environmental impact assessment and environmental impact assessment report.*
- *Public opinions and suggestions shall be fully evaluated and considered by the competent authority for environmental protection and relevant industrial sectors and forwarded to the construction units.*

Various principal provisions concerning public participation in environmental impact assessment were stipulated in Article 13 of the Water Pollution Law (1996) and Article 13 of the Noise Pollution Law (1996) to ensure that the opinions of relevant organizations and local residents at the sites of construction projects shall be included in the environmental impact assessment reports. Furthermore, as stated in the *Ordinances of Management for Environmental Protection of Construction Project*, promulgated by the State Council in November 1998, the construction unit shall solicit the opinions of the relevant organizations and local residents at the sites of construction projects according to the relevant laws and regulations, while compiling the environmental impact assessment reports.

As regulated in Article 58 of *the Law of the People's Republic of China* on *Legislation* (the Legislation Law) promulgated in March 2000, during the drafting of administrative regulations, the opinions of relevant agencies, organizations, and the public shall be solicited through forums, symposiums, consultation meetings, hearings, and so on. Several provisions were further specified in the EIA Law, promulgated in October 2002, as shown in Box 12.2.

Box 12.2 Provisions of public participation specified in the EIA Law (2002)
- *Article 5: Relevant organizations, experts and the public are encouraged to participate in environmental impact assessment in suitable channels.*
- *Article 11: Prior to submitting the draft special plan for review and approval, the agency who compiles the special plan shall solicit the opinions and comments of relevant organizations, experts and the public on the draft environmental impact assessment reports, through symposiums, consultation meetings, hearings, or other forms, as the special plan might cause adverse environmental impacts which will directly affect public environmental rights and interests; and these opinions and comments shall be*

> *seriously considered, and shall be included in the environmental impact assessment reports to explain whether these opinions and comments were adopted or not, and why.*
> - *Article 21: Except for classified information regulated by the State, for construction projects with potential significant environmental impacts to be required to prepare environmental impact assessment report, the construction unit shall solicit the opinions and comments of relevant organizations, experts and the public, through symposiums, consultation meetings, hearings, or other forms, prior to submitting the environmental impact assessment reports for review and approval; and these opinions and comments shall be seriously considered, and shall be included in the environmental impact assessment reports to explain whether these opinions and comments were adopted or not, and why.*

Apparently, as required and protected by law, public participation in environmental management in China has made great progress in legislation.

The public hearing of environmental impact assessment of the lake bed seepage-proofing project in the Yuan-Ming-Yuan Imperial Garden (the Old Summer Palace), held in April 2005, is widely considered a landmark of public participation in the area of environmental protection in China (Ma, Webber, & Finlayson, 2009). Since then, the MEP has explored several methods and attempts to promote public participation, and many provisions concerning public participation were explicitly stipulated in the relevant laws and regulations, as shown in Box 12.3.

Box 12.3 Provisions concerning public participation in various laws and regulations
- the *Interim Measures of Public Participation in Environmental Impact Assessment (February 2006).*
- the *Measures of Environmental Information Disclosure (on Trial) (February 2007).*
- the *Guiding Opinions Concerning Cultivating and Regulating the Orderly Development of Environmental Non-Governmental Organizations (December 2010).*
- the *Guidelines for Disclosure of Government Information on EIA for Construction Project (on Trial) (November 2013).*
- the *Guiding Opinions Concerning Promoting Public Participation in Environmental Protection (May 2014).*
- the *Measures of Public Participation in Environmental Protection (July 2015).*
- the *Measures of Public Participation in Environmental Impact Assessment (July 2018).*

As explicitly required by the CPC and the State, the promotion of the public participation in the field of environmental protection legally and orderly is an objective necessity for speeding up the transformation of economic and social development, and comprehensively strengthening reforms. Per the report of the 18th NCCPC, to safeguard people's Right to Know, Right to Participate, Right of Free Expression, and Right of Supervision is to guarantee the correct operation of power. In the EP Law (2014), the principle of public participation was clearly stipulated in the General Regulations, and a special chapter to regulate information disclosure and public participation was provided. Moreover, in the *Opinions on Accelerating the Advancement of the Ecological Civilization Construction*, promulgated in May 2015 by the CCCPC and the State Council, the following goals were explicitly stipulated: to encourage active public participation; to perfect the public participation system; to precisely disclose various kinds of environmental-related information in a timely manner and to broaden the extent of information to be disclosed so as to ensure the public's Right to Know and to safeguard the public's environmental rights and interests. In July 2015, the *Measures of Public Participation in Environmental Protection* was issued by the MEP, making it the first departmental regulation to include a specific provision for public participation in the field of environmental protection since the promulgation of the EP Law (2014) (Wu, Chang, Yilihamu, & Zhou, 2017). In July 2018, with the promulgation of the *Measures of Public Participation in Environmental Impact Assessment* by the MEE, public participation has become the indispensable part of the EIA.

12.2 Channels of Public Participation in Environmental Management

By soliciting comments, conducting surveys, organizing symposiums, convening expert demonstration meetings and hearings, and so on, the competent authority for environmental protection shall seek opinions and suggestions on environmental protection-related matters or activities from citizens, legal persons and other organizations. In addition, the public's opinions and suggestions can be conveyed to the competent authority for environmental protection through phone, fax and internet (Chang & Wu, 2011).

(1) Questionnaire survey

In order to seek opinions and suggestions through questionnaires, the competent authority for environmental protection shall disclose and explain the information concerning the relevant matters or activities. Moreover, the questions in the questionnaire shall be simple and explicit enough for easy comprehension by the general public. The number and range of participants shall be determined by a comprehensive consideration of the extent and degree of the environmental impact of the relevant matters or activities, the degree of social concerns, the human and material resources required for the organization to participate, and so forth.

(2) Workshop and consultation meeting

In order to seek opinions and suggestions through questionnaires, the competent authority for environmental protection shall disclose and explain the information concerning the relevant matters or activities. Moreover, the questions in the questionnaire shall be simple and explicit enough for easy comprehension by the general public. The number and range of participants shall be determined by a comprehensive consideration of the extent and degree of the environmental impact of the relevant matters or activities, the degree of social concerns, the human and material resources required for the organization to participate, and so forth.

(3) Public hearing

In case the matter is subject to a public hearing according to the relevant laws and regulations, the competent authority for environmental protection shall publicize the matter and summon a public hearing in accordance with the principles of openness, fairness, impartiality, and convenience. In addition, the opinions and suggestions of citizens, legal persons, and other organizations shall be carefully considered and the public's rights to express, inquire, and appeal shall be ensured. Except for classified information concerning state secrets, trade secrets, or personal privacy, the public hearing should be held publicly. The opinions and suggestions of citizens, legal persons, and other organizations shall be carefully categorized, analyzed, and considered by the competent authority for environmental protection prior to any decision-making on environmental issues. Feedback shall be provided to citizens, legal persons, and other organizations in an appropriate manner.

(4) Public supervision

Public supervision of environmental affairs of citizens, legal persons, and other organizations shall be strongly supported and encouraged by the competent authority for environmental protection. In case of environmental pollution and ecological destruction caused by any units and individuals, any citizens, legal persons, and other organizations shall report to the competent authority for environmental protection through various channels, including letters, fax, email, environmental protection hotline "12369", and government websites.

(5) Citizen impeachment

In case of any dereliction of duty or breach of duty on the part of local governments at all levels or the competent authority for environmental protection at or above the county level, any citizens, legal persons and other organizations shall have the right to report to the superior authorities or supervisory institutes. In accordance with the relevant laws and regulations, the competent authority for environmental protection at the higher level receiving the report shall investigate and verify the reported matters, and inform the informant of the results of the investigation and its handling. Moreover, the competent authority for environmental protection at the higher level receiving the report shall keep the relevant information of the informant confidential and protect the lawful rights and interests of the informant.

(6) Enforcement

The competent authority for environmental protection shall provide legal consultation, written comments, and assistance in the processes of investigation and evidence collection to support the environmental public interest litigation raised by environmental NGOs according to the relevant laws. In addition, within the scope of duty, the competent authority for environmental protection shall strengthen the propaganda and education work to (a) popularize knowledge of environmental science and enhance public awareness of environmental protection and resources conservation, (b) encourage the practice of green living and green consumption so as to customize the social ethics and public morals of a low-carbon economy, resources conservation, and environmental protection. Moreover, the competent authority for environmental protection may support and guide the environmental NGOs to participate in environmental protection activities through various means, such as financing projects and purchasing services.

12.3 Other Means of Public Participation in Environmental Management

Currently in China, the public may participate in environmental management through the following indirect ways: the People's Congress, the Political Consultative Conference, the democratic parties, the local government departments to supervise and manage environment protection, the People's Courts, residents' committees and village committees, the staff and workers' representative congress, public opinion, and so on.

(1) People's Congress at all levels

Through the People's Congress, the public can participate in environmental management in two different ways. The first channel is the legislative process of the People's Congress in which the representative of the People's Congress shall have the right to propose or examine the propositions related to environmental protection, during the meeting of the People's Congress. The second way is to exercise the constitutional right of supervision by the People's Congress on the administrative activities of the government in which the representative of the People's Congress shall have the right to supervise the environmental protection work of the government, no matter whether it is in session or not, through the following activities, as shown in Box 12.4.

> **Box 12.4 Major schemes of the representative of the People's Congress to supervise the environmental protection work of the government**
> - *to review and deliberate the government's report on environmental protec-tion and make a decision.*
> - *to propose the proposition related to environmental protection during the meeting of the People's Congress, and then submit the proposition to relevant government departments and environmental protection agencies whom shall then respond to the proposition by submitting a report on the proposition to the People's Congress; and this report shall be forwarded to the representative proposed the proposition.*
> - *to address inquires to the government departments or agencies whom shall then respond to the inquiries.*
> - *to make decision on specific environmental affairs: Local People's Congress shall have the right to make decision on local environmental issues and to order local governments or other organizations to eradicate the environmental problems.*
> - *to inspect special environmental affairs in secret, which means no advanced notice or announcement prior to the inspection.*
> - *to criticize and comment on the environmental protection work of the government so as to coerce the government to pay more attention on the environmental affairs pointed out by the representative(s) during the plenary session or group meeting of the People's Congress.*

(2) Chinese People's Political Consultative Conference at all levels

The Chinese People's Political Consultative Conference (the CPPCC) is the basic form of cooperation between political parties and political consultation system under the leadership of the CPC. In the Constitution, the role of the CPPCC in the political, economic and social development of the country was affirmed, and the Right of Free Expression of the CPPCC members during the CPPCC meeting at all levels was ensured. The public can express their demands to the CPPCC members at all levels. As the CPPCC members have the rights to recommend, inspect, and report, the relevant departments shall carefully study the reports and recommendations of the CPPCC members and actively prepare corresponding responses. Actually, in China, the CPPCC at all levels very actively participate in environmental management, such as conducting organized inspections and investigations into environmental problems and providing many insightful and valuable suggestions and criticisms on the CPC and the governments at all levels. These suggestions and criticisms play an important role in improving the decision-making of the CPC and the governments on environmental affairs.

12.3 Other Means of Public Participation in Environmental Management

(3) Democratic parties

Right now, in China, there are eight democratic parties entitled to investigate environmental affairs and submit reports and suggestions to the CPPCC or the CPC and the government, directly. There are several means of cooperation and collaboration between the CCCPC and eight democratic parties, as listed in Box 12.5. Through these channels, the views and recommendations of the democratic parties on environmental affairs can directly permeate the decision-making process of the government so as to improve the environmental decision-making.

> **Box 12.5 Major means of cooperation and collaboration between the CCCPC and democratic parties**
> - *Annual meetings between the CCCPC and the leaders of the democratic parties and nonpartisans.*
> - *Bimonthly meetings (twice a month) will be convened by the CCCPC to annunciate some important information, to solicit opinions of democratic parties and nonpartisans, or to exchange viewpoints with democratic parties and nonpartisans.*
> - *Unscheduled forums.*
> - *Regarding important matters, democratic parties or nonpartisans shall submit written recommendations to the CCCPC or invite the leaders of the CPC for further discussion.*

(4) Local government departments to supervise and manage environment protection

The public have the right to make accusations against or put forward some requirements for the government departments in charge of supervising and managing environment protection concerning environmental affairs. The government departments in charge of supervising and managing environment protection, including the competent authority for environmental protection, the competent authority for public health, and other relevant departments, are obliged to pay attention to the public's accusations or suggestions on environmental affairs. As the leading department for environmental protection supervision and management, the competent authority for environmental protection shall be primarily responsible for paying attention to the public's opinions and providing appropriate responses. In order to maintain a close relationship with the public and to enhance the efficiency of public services, the procedures to handle the public's accusations and requirements, in a transparent manner, have been established by the competent authority for environmental protection and local governments.

(5) People's court at all levels

The public can approach the local people's court at all levels to file a lawsuit against the polluters. The people's court shall judge the lawsuit according to the *Civil Procedural Law of the People's Republic of China* and the EP Law. A judicial decision against a single polluter generates a frightening and deterring effect on similar polluters. In this sense, as the most direct way to participate in environmental management, an environmental lawsuit filed by an individual promotes national environmental management.

(6) Residents' committees and village committees

Residents' committees in urban areas and village committees in rural areas are grassroots-level self-governing organizations. The directors, deputy directors, and members of the committees are elected by the residents/villagers and shall be responsible for the environmental affairs in the streets or villages. Thus, residents/villagers can convey their opinions on environmental affairs to these committees.

(7) Public opinion

As a means and pressure to promote environmental protection, public opinion plays an essential role in the work of environmental protection and management. Usually, the relation between the government and the public is determined by the effectiveness of environmental protection and management. If environmental pollution and ecological degradation cannot be effectively controlled, the government will lose the public's trust. Therefore, the government takes measures to disclose environmental information so as to encourage the public to take the initiative in supervising the environmental management activities.

12.4 Environmental Information Disclosure

The concept of information disclosure first appeared in the *Freedom of Information Act* (United States, 1966). Triggered by the Bhopal disaster in India in December 1984, the Emergency Planning and Community Right-to-Know Act (EPCRA) was promulgated in the United States in 1986 in response to concerns regarding the environmental and safety hazards posed by the storage and handling of toxic chemicals, in which government departments are required to disclose information of plant pollutants to the public, and companies are required to release information concerning the amount of more than 600 different kinds of toxic chemicals to the public (Graham & Miller, 2001).

In the 1990s, environmental information disclosure was listed on the agenda of global environmental policy. Many developed countries also actively encouraged their governments and companies to disclose information to the public. In June 1998, the *Convention on Access to Information, Public Participation in Decision-making and Access to Justice in the Environmental Matters* (Aarhus Convention) was approved by the United Nations Economic Commission for Europe (UNECE) during the fourth ministerial meeting. In addition to providing detailed specifications of the environmental information disclosure system, the Aarhus Convention clearly defined the basic concept of "environmental information" and "public authority", explicitly regulated the main body, content, exceptions, and judicial remedies mechanism of

12.4 Environmental Information Disclosure

government environmental information disclosure; distinctly stipulated the principles and the implementing schemes of corporate environmental information disclosure and product environmental information disclosure; and specifically clarified the perfection and development mechanism of the environmental information disclosure system. The signing of the Aarhus Convention has resulted in huge international repercussions.

After the Reform and Opening-Up, the CPC and the Central Government gradually began paying attention to the information disclosure system. Since the 1980s, in order to meet the requirements of reform and opening-up and the development of the socialist market economy, the Chinese government has been continuously promoting and deepening the reform of the administrative system. In particular, the openness of government affairs is one of the important goals of the reform (Wu et al., 2017).

In April 2007, the *Ordinances of the People's Republic of China on Government Information Disclosure* was promulgated by the State Council to ensure the right of the citizens, legal persons, and other organizations to acquire government information, to enhance the transparency of government work, to promote the rule of law, and to provide a legal guarantee for fulfilling the service functions of government information to the production, living, economic, and social activities of the public. Subsequently, in April of the same year, the Measures of Environmental Information Disclosure (on Trial) was enacted by the SEPA to further promote and regulate the competent authorities for environmental protection and companies to disclose the environmental information, to safeguard the rights of the citizens, legal persons, and other organizations to acquire environmental information, and to promote public participation in environmental protection.

(1) Government environmental affairs information disclosure

Since the promulgation of the *Ordinances of the People's Republic of China on Government Information Disclosure*, through continuous efforts in constructing and enhancing government websites at all levels, all kinds of government environmental affairs information have been unveiled to the public. Moreover, annual performance reviews on the websites of provincial environmental protection bureaus shall be carried out by the MEP. The reports of the performance evaluation are to be publicized. Right now, the functions of information disclosure, online services, and interactions between the government and the public can be basically fulfilled on all the websites of the competent authorities for environmental protection above the provincial level.

(2) Environmental quality information disclosure

Air Quality: Currently in China, the air quality monitoring data is relatively comprehensive and timely, as there are 1436 air quality monitoring stations allocated in 338 cities. The real-time air quality information, including particulate matter (PM_{10}), fine particulate matter ($PM_{2.5}$), sulfur dioxide (SO_2), nitrogen dioxide (NO_2), ozone (O_3), and carbon monoxide (CO), as well as the air quality index (AQI), is disclosed to the public through the Platform of National Real-Time Urban Air Quality Data.

Water Quality: Right now in China, more than 100 water quality automatic monitoring stations have been installed to monitor the water quality of 63 rivers and 13 lakes and reservoirs, which cover the seven key river systems. Water quality indexes include water temperature, pH, dissolved oxygen, conductivity, turbidity, permanganate index, total organic carbon, and ammonia nitrogen. In addition, total nitrogen and total phosphorus are monitored for lake water quality. Surface water quality information is publicized through the System of National Real-Time Automatic Monitoring Surface Water Quality Data.

Soil Quality: Soil environmental quality information disclosure is still in the initial phase. So far, there are no websites or platforms built by the relevant government departments to disclose soil quality information. The first nationwide survey on the status quo of soil contamination was performed from April 2005 to December 2013, to cover an area of 6.3×10^6 km^2. For more comprehensive information, please refer to the *Bulletin of the First Nationwide Survey on Soil Contamination Status Quo*, published by the MEP and the MLR in April 2014.

Solid Waste: Solid waste information disclosure is gradually being standardized. As stipulated in Article 12 in the Solid Waste Pollution Law, the information concerning the categories, amount, and disposal status quo of solid waste shall be publicized regularly by the competent authority for environmental protection of the people's governments in large and medium-sized cities; and the same type of information provided in the previous year shall be released by June 5 every year. Through these years, the number of cities that publish relevant information has been continuously increasing, and the content of disclosed information has been gradually enriched.

(3) Pollution sources information disclosure

In July 2013, the *Notice on Strengthening the Information Disclosure of Environmental Supervision on Pollution Sources* was promulgated by the MEP to stress that the competent authority for environmental protection at all levels shall proactively disclose information about environmental supervision on pollution sources, according to the requirements of the Appendix 1, *the Catalogue of Information Disclosure of Environmental Supervision on Pollution Sources (the First Batch)*, of *the Notice on Strengthening the Information Disclosure of Environmental Supervision on Pollution Sources*. In order to further encourage enterprises and public institutions to disclose environmental information, the *Interim Measures of Enterprises and Public Institutions to Disclose Environmental Information* was enacted by the MEP in December 2014 to regulate that enterprises and public institutions must disclose environmental information and to explicitly specify the content, method, time limit of disclosure and legal responsibilities of failure to disclose information.

In China, except for the online real-time monitoring information, in general, the development of environmental information disclosure on pollution sources has achieved some progress. Some companies have complied with mandatory requirements of information disclosure. Besides, some large enterprises voluntarily disclose information on corporate environmental behaviors through publishing of environmental performance reports so as to fulfill their corporate social responsibilities and elevate their corporate image.

(4) Information disclosure in EIA

After the effectiveness of *the Interim Measures of Public Participation in Environmental Impact Assessment* in March 2006, the construction project and the EIA consulting services agency were required to release the abridged version of EIA reports to the public. From 2006 to 2013, it was common for EIA consulting services agencies to only disclose basic information and the abridged reports on their own websites. Since the issuance (November 2013) and implementation (January 2014) of the *Guidelines for Disclosure of Government Information on Environmental Impact Assessment for Construction Project (on Trial)*, the public have access to the full version of EIA reports from the website of the competent authority for environmental protection. The evolutionary process of information disclosure in EIA in China has been thoroughly discussed in Sect. 4.5.4, as shown in Fig. 4.2.

References

Chang, I.-S., & Wu, J. (2011). Planning and rationalization of public participation in China's environmental management. *Management Science & Engineering, 5*(1), 37–50.

Graham, M., & Miller, C. (2001). Disclosure of toxic releases in the United States. *Environment Science & Policy for Sustainable Development, 43*(8), 8–20.

Ma, J., Webber, M., & Finlayson, B. L. (2009). On sealing a lakebed: mass media and environmental democratisation in China. *Environmental Science & Policy, 12*(1), 71–83. https://doi.org/10.1016/j.envsci.2008.09.001. [2019-10-08].

Wu, J., Chang, I. Shin, Yilihamu, Q., & Zhou, Y. (2017). Study on the practice of public participation in environmental impact assessment by environmental non-governmental organizations in China. *Renewable and Sustainable Energy Reviews, 74,* 186–200.

Chapter 13
Restraining Redlines System of Ecological and Environmental Protection

In China, the notion of demarcating the ecological redline was first proposed in October 2011. Then, the ecological redline has been gradually elevated as one of the government's primary tasks of ecological protection and ecological civilization construction. Currently in China, the concept of ecological redline has been widely applied to the fields of resources conservation and ecological protection and management and gradually developed into the restraining redlines system, including the energy consumption redline, water resource redline, cultivated land redline, and ecological protection redline, to impose serious spatial and quantitative limits on resource utilization and environmental damage (He et al., 2018).

13.1 The Initiation and Development

In October 2011, the *Opinions Concerning Strengthening the Key Works of Environmental Protection* was promulgated by the State Council to propose the concept of environmental functional zoning nationwide, and the demarcation of ecological redlines in key ecological function zones, and in ecological sensitive and fragile zones of terrestrial and marines areas, to denote the initiation of the ecological redline.

In March 2012, the National Symposium on Demarcation Technologies of Ecological Redline was convened by the MEP to summon domestic famous experts and the leading officials of the competent authorities for environmental protection of major provinces to thoroughly deliberate and communicate the concept, connotation, and demarcation technologies and methods of ecological redline, as well as to generally deploy the national demarcation work of ecological redline. By the end of 2012, Inner Mongolia and Jiangxi were designated as the demonstration sites for the pilot program of ecological redline.

In order to construct ecological civilization, many schemes were proposed during the third plenary session of the 18th CCCPC held in November 2013, as listed in Box 13.1. In order to fulfill the *Opinions Concerning Strengthening the Key Works of Environmental Protection* and the directive of the third plenary session of the 18th

CCCPC, the *National Ecological Redline—the Technical Guidelines for the Demarcation of Ecological Function Zone Redline (on Trial)* was publicized in January 2014 to further guide the work of demarcating ecological redline, to ensure national ecological safety, and to promote economic and social sustainable development. As stipulated in Article 29 of the EP Law (2014), revised in April 2014, the government shall demarcate ecological redlines in key ecological function zones, and in ecological sensitive and fragile zones of terrestrial and marines areas, so as to implement strict ecological protection.

Box 13.1 Key schemes proposed by during the third plenary session of the 18th CCCPC

- *to establish a systematic and complete ecological civilization system so as to employ the system to protect the ecological environment.*
- *to improve the property right system and utilization control system of natural resources.*
- *to demarcate the ecological protection redlines.*
- *to implement the paid use system of resources and the ecological compensation system.*
- *to reform the ecological environment protection and management system.*

As stated in *the Opinions on Accelerating the Advancement of the Ecological Civilization Construction* promulgated in April 2015, the ecological redline, consisting of "the upper limit of resource consumption", "the bottom line of environmental quality", and "the ecological protection redline", shall be stringently abided by. At the end of April 2015, the *Technical Guidelines for the Demarcation of Ecological Redline* was released by the MEP to further clarify the concept and characteristics of the ecological redline, to propose the technique of demarcating the ecological redline, to guide the demarcation of the ecological redline nationwide, and to safeguard national and regional ecological security. In September of the same year, the *General Planning for the Institutional Reform of Ecological Civilization* was issued by the CCCPC and the State Council, jointly, to propose the important measure of demarcating the production space, living space, and ecological space. The ecological redline is the bottom line of national and regional ecological security. In order to develop the structure of three-space and three-redline, through the integration of multiple plans, the ecological redline and the ecological space shall be firstly determined to form the basis of national land space planning so as to rationally arrange the production space and the living space and to define the urban expansion boundary and permanent farmlands.

In March 2016, the *Outlines of the Thirteenth Five-Year Plan for the National Economic and Social Development of the People's Republic of China* was passed by the fourth session of the 12th NPC to specify some requirements in Chapter 47, "Perfecting the Safeguarding System of Ecological Security", such as realizing the regulation and control of the utilizations of ecological space, demarcating and safeguarding the ecological redline, and ensuring the functions, areas, and essence of the ecological function zone. In February 2017, the *Opinions Concerning Demarcating and Safeguarding the Ecological Redline* was jointly issued by the General Office of the CCCPC and the General Office of the State Council to explicitly propose a work schedule of demarcating the ecological redline, as shown in Box 13.2.

Box 13.2 Work schedule of demarcating ecological redline

- *to demarcate ecological redline in Beijing-Tianjin-Hebei region, and in the provinces (municipality) along the Yangtze River Watershed, by the end of 2017.*
- *to demarcate ecological redline in the rest provinces (municipality), by the end of 2018.*
- *to complete the survey and demarcation of ecological redline, nationwide, so as to establish the ecological redline system, to effectively optimize and protect the ecological space, nationwide, to stabilize the ecological functions, and to perfect the structure of national ecological security, by the end of 2020.*
- *to further optimize the layout of ecological redline, to effectively implement the ecological redline system, to significantly elevate the ecological functions, and to fully safeguard the national ecological security, by the end of 2030.*

The ecological redline strategy covers many important ecological function zones (water and soil conservation, biodiversity preservation, windbreak, sand-fixation, and so on), and ecologically fragile regions that are prone to soil erosion, desertification and salinization. Therefore, in the context of a fragile ecological environment and crucial condition of ecological security, the ecological redline serves as both the bottom line and lifeline in safeguarding national ecological security (Xinhua, 2017).

13.2 Practices of Redline in Resources and Environment

13.2.1 Cultivated Land Redline

The connotation of cultivated land redline includes two parts, the total area of cultivated land and the area of permanent cultivated land, which can never be overstepped. However, due to the needs of development, non-permanent cultivated land might be occupied and transferred into construction land on condition that the occupied and transferred non-permanent cultivated land shall be recompensed through barren land reclamation, off-site compensation, and other effective means.

In March 2006, the *Outlines of the Eleventh Five-Year Plan for the National Economic and Social Development of the People's Republic of China* was approved by the fourth session of the 10th NPC to unambiguously stipulate that the total area of cultivated land shall be conserved for no less than 1.8×10^9 Mu (1.2×10^6 km^2, at 1 km^2 = 1500 Mu), which is the cultivated land redline that serves as the legally binding indicator for the next five years,. In October 2008, *the Outlines of the General Planning for National Land Utilizations (2006–2020)* was issued by the State Council to specify six binding indicators and nine anticipated indicators while setting the objectives and tasks of land utilization, and to further affirm the 1.8×10^9 Mu as the cultivated land redline.

13.2.2 Water Resource Redline

In December 2010, the *Decisions on Accelerating the Reform and Development of Water Conservancy*, issued by the CCCPC and the State Council, defined the three redlines to strictly regulate the water resource management system, as shown in Box 13.3. The three redlines of water resource management are focused on the sustainable utilization of water resources, and the coordination between social and economic development and resources conservation, which play an essential role in transforming the mode of water management, promoting the construction of a water conservation society, and improving water ecological environment.

Box 13.3 Three redlines for water resources

- *The first redline is to control water resources development and utilization. By 2020, the total annual water consumption shall be confined to 6.70×10^{11} m^3, which is a binding indicator.*
- *The second redline is to control water use efficiency. By 2020, the effective utilization coefficient of farmland irrigation water must be elevated above 0.55, and the amount of water consumption per 10,000 yuan Gross Domestic Product (GDP) and 10,000 yuan Industrial Value-Added (IVA) shall be reduced to 124 m^3/¥10,000 GDP and 60 m^3/¥10,000 IVA, respectively.*
- *The third redline is to limit the amount of pollutants in water function zones. The attainment rate of water function zones in major rivers and lakes shall be elevated to 60%.*

In February 2012, the *Opinions on Implementing the Strictest Water Resource Management System* was issued by the State Council to clearly put forward four systems (the system of total water use control, the system of water use efficiency control, the system of limiting pollutants discharge in water function zones, and the system of liability and performance evaluation) and three redlines as the fundamental

13.2.3 Emission Cap Redline

The emission cap redline is the limit of pollutant emission within a certain region, in order to meet the requirements of regional environmental quality. The emission cap redline is the typical redline implemented in the environmental management system, which was initiated during the 9th FYP, but concretely realized during the 11th FYP. The emission cap was first applied to COD and SO_2 during the 11th FYP by allocating the allowance to various cities and pollution units. To establish the liability of noncompliance with the emission cap redline, limited approval for new emission applications shall be imposed to nonattainment areas or enterprises. In the context of emission cap redline, the total emissions of COD and SO_2 in 2010 dropped by 14.3% and 12.5%, respectively, in comparison with that of 2005, achieving the target of five-year environmental protection plan for the first time.

Many stringent schemes of emission cap redlines, such as "limited regional approval", "limited industrial approval", "suspending review and approval process", "disclosure mechanism", "pollution governance within certain timeframe", and "penalty", were explicitly stipulated in various environmental protection laws and regulations, and plans for energy conservation and emission reduction, introduced by the State Council since 2007, and the Water Pollution Law revised in 2008, to ensure the concrete execution of the emission cap redline. Since the beginning of the 12th FYP, the coverage of the emission cap redline was expanded from two pollutants (SO_2, COD) to four pollutants (SO_2, COD, NO_x, NH_3–N), in addition to heavy metal pollution control (see Sect. 10.1).

13.3 Framework Design for Redline in Resources and Environment

The ecological redline, comprised of the "upper limit of resource consumption", "the bottom line of environmental quality", and "the ecological protection redline", is to enhance the binding indicators of resources conservation and environmental protection, and to restrain all kinds of economic and social activities within the scope of the redline, as shown in Fig. 13.1.

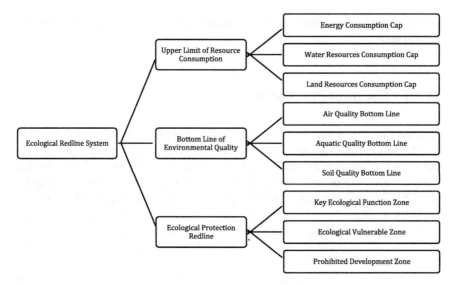

Fig. 13.1 Composition of ecological redline system

13.3.1 Upper Limit of Resource Consumption (Resource Consumption Cap)

The upper limit of resource consumption (including energy consumption cap, water resources consumption cap, and land resources consumption cap) is the ceiling imposed on resource consumption to promote energy and resource conservation, and to ensure the high efficient utilizations of resources (such as energy, water, and land), provided that the basic needs of economic and social development shall be guaranteed, and the current carrying capacity of resources and environment shall be satisfied.

(1) Energy consumption cap

The energy consumption cap is the objective of total energy consumption control determined on the basis of level of economic and social development, industrial structure and layout, resource endowment, environmental capacity, emission reduction requirements and other factors.

(2) Water resources consumption cap

The water resources consumption cap is the objective of total water resources consumption control established according to water resource endowment, ecological water demand, and the rational demand from economic and social development. In areas with serious water shortage and groundwater overdraft, the ceiling of groundwater exploitation, ecological water demand, and ecological flow shall be strictly regulated.

(3) Land resource consumption cap

The land resource consumption cap is the objective of total land resources consumption control determined according to food security, ecological security, major function settings, development intensity, urban and rural population scale, and per capita construction land requirements, in which the total area of cultivated land shall be determined, the area of permanent cultivated land demarcated, and the requisition-compensation balance of cultivated land implemented. In areas with prominent supply and demand imbalance of land use, the objective of total urban and rural construction land control shall be set up.

13.3.2 Bottom Line of Environmental Quality (Environmental Quality Baseline)

The core of the bottom line of environmental quality is to improve environmental quality so as to guarantee environmental and public health, and people's life and property safety. The bottom line of environmental quality is comprised of the air quality baseline, water quality baseline, and soil quality baseline, which are determined on the basis of environmental quality status quo, the needs of economic and social development, the advancement of pollution prevention and control technologies, and the regional objectives of water quality, air quality, and soil quality, at different stages.

(1) Water quality baseline

In order to be in line with the Action Plan for Water Pollution, the water quality baseline is meant to demarcate the control units of water quality, and to explicitly stipulate the requirements of water quality improvement, pollutant emission control, and risk management, according to the evaluation of the importance, sensitivity, and vulnerability of the functions of water environment.

(2) Air quality baseline

In order to be in compliance with the *Ambient Air Quality Standards (2012)* and to be in line with the Action Plan for Air Pollution, the air quality baseline is meant to demarcate the control units of air quality, and to explicitly stipulate the requirements of air quality improvement, pollutant emission control, and risk management, according to the evaluation of the importance, sensitivity, and vulnerability of the functions of air environment.

(3) Soil quality baseline

In order to be in line with the Action Plan for Soil Pollution, the soil quality baseline is meant to demarcate the control units of soil quality, and to explicitly stipulate the requirements of pollutant emission control and risk management, according to the evaluation of the importance, sensitivity, and vulnerability of the functions of soil environment.

13.3.3 Ecological Protection Redline

The ecological protection redline is designated to protect the higher graded areas in key ecological protection zones identified through the evaluation of the significance of ecosystem services and the assessment of ecological sensitivity and vulnerability, according to the EP Law and the relevant normative documents and technical methods of ecological protection. Specifically, it includes the redline of the key ecological function zones protection, the redline of the ecological sensitive and vulnerable zones protection, and the redline of the prohibited development zones protection.

(1) Key ecological function zones protection redline

The redline of key ecological function zones protection is to protect the higher graded areas in key ecological function zones determined through the evaluation of the significance of ecosystem services, including water source protection, soil and water conservation, windbreak and sand fixation, and biodiversity preservation.

(2) Ecological sensitive and vulnerable zones protection redline

The redline of ecological sensitive and vulnerable zones protection is meant to protect the higher graded areas in ecological sensitive and vulnerable zones categorized through the assessment of ecological sensitivity and vulnerability, including soil and water erosion, desertification, and hamada.

(3) Prohibited development zones protection redline

The redline of prohibited development zones protection is meant to protect the natural reserves. In principle, the entire natural reserve shall be included in the redline. However, for a larger natural reserve, the core zone and the buffer zone, of course, shall be included in the redline; however, the extent of the transition zone to be included in the redline shall be determined through the evaluation of the significance of ecological protection in combination with regionalized management (Jiang, Bai, Wong, Xu, & Alatalo, 2019).

References

He, P., Gao, J., Zhang, W., Rao, S., Zou, C., Du, J., et al. (2018). China integrating conservation areas into red lines for stricter and unified management. *Land Use Policy, 71*, 245–248. https://doi.org/10.1016/j.landusepol.2017.11.057. [2019-10-17].

References

Jiang, B., Bai, Y., Wong, C. P., Xu, X., & Alatalo, J. M. (2019). China's ecological civilization program–Implementing ecological redline policy. *Land Use Policy, 81,* 111–114. https://doi.org/10.1016/j.landusepol.2018.10.031. [2019-10-17].

Xinhua. (2017, Feb 7). China to complete drawing ecological 'red line' by 2030. Retrieved March 21, 2019, from http://english.gov.cn/policies/latest_releases/2017/02/07/content_281475561612035.htm. [2019-03-21].

Chapter 14
Compensation for Environmental Damage

In the legal sense, environmental damage is comprised of two levels. The first is private rights damage, such as damage to personal health and property rights caused by man-made pollution. The second is an adverse change in the environmental elements (including atmosphere, groundwater, surface water, and soil) and the biological elements (such as plants, animals and microorganisms), as well as the functional degradation of the ecosystem consisting of these elements, due to the environmental pollution and ecological destruction (Faure & Liu, 2012).

As for private rights damage, personal health and property rights are the objects of civil rights in Private Law and are subject to the protection and remedies in Civil Law. As stipulated in Article 65 of the *Law of the People's Republic of China on Tort Liability* (the Tort Liability Law), a polluter shall be liable for tort in case of damage caused by environmental pollution. In Article 64 of the EP Law, it was regulated that any damage resulting from environmental pollution and ecological destruction shall be subject to tort liability, in accordance with the relevant provisions of the Tort Liability Law. In a lawsuit against the infringement of property rights and personal rights resulting from environmental pollution, the main body of the plaintiff can only be citizens and legal persons whose rights were infringed. In other words, the plaintiff must be the victim of infringement, and the state administrative agencies have no right to bring civil litigation on behalf of any person.

For the latter, namely, environmental damage, owing to the fact that the object of damage is the environmental public interest carried out by the ecological environment and does not qualify for any attribution of personal and property rights, legal rights in the Private Law lack clarity and definition, and no specific subjects of rights infringement exist according to the Private Law (Faure & Liu, 2014). Therefore, compensation for environmental damage is proposed to address this problem by revising and adjusting the conventional tort theory to serve the purpose of safeguarding public interests and promoting environmental protection under the new circumstances.

© Chemical Industry Press and Springer Nature Singapore Pte Ltd. 2020

14.1 International Experience

The practices of compensation for environmental damage have existed in the United States and the European Union for several decades and have helped to gradually establish relatively complete legal systems and execution mechanisms (Wilde, 2013). The most prominent feature of these practices is that specialized legislation was introduced to be differentiated from that for personal property damage, due to the special peculiarity of environmental damage. Besides, remedies in different laws shall be applied to personal property damage and environmental damage.

14.1.1 In the United States

In the United States, the scope of compensation for natural resources damage includes pollution cleanup costs, pollution remediation costs, and the loss of ecological services during the period of remediation and restoration, as well as the assessment costs, which shall all be borne by the polluter. As a remedial measure for natural resources damage, the environmental public interest litigation system was established to protect public environmental interests. The environmental public interest litigation can be resorted to in case of direct or indirect damage to ecological functions, environmental aesthetics, and other general public interests, wherein damage to personal or property rights is not a precondition. The competent authority for environmental protection, social organizations, and individual citizen are all eligible, as the plaintiff, to bring litigation against enterprises for their pollution emission behaviors.

It should be pointed out that American judicial practices tend to reduce the amount of litigation through negotiation. For example, in the environmental damage claims, prior to any formal allegation against the responsible party brought by the Environmental Protection Agency (EPA) and other government agencies, an agreement on environmental compensation, restoration, and other matters can be reached with the responsible party. The responsible party shall compensate and restore the environmental damage in accordance with the agreement. If an agreement cannot be reached, the EPA may restore the environment on behalf of the responsible party, and then, a lawsuit will be filed to require the responsible party to reimburse the cost of environmental compensation and restoration.

In the United States, there are two sets of rules to appraise the natural resource damage performed by the Department of the Interior and the National Oceanic and Atmospheric Administration (NOAA), separately, to include the following major steps and methods, as shown in Box 14.1 (Kubasek & Silverman, 2014).

> **Box 14.1 Major steps and methods of natural resource damage assessment in America**
> - *to determine the nature, extent and scope of environmental damage and to confirm the environmental baseline before the damage occurred.*
> - *to determine the causal relationship between pollution behaviors and the damage.*
> - *to quantify the environmental damage through equivalency analysis.*
> - *to determine the scale and degree of basic restoration in order to restore the damaged environment to the baseline state.*
> - *to determine the scale and degree of compensation so as to make up losses during the period.*
> - *to determine the scale and degree of additional restoration measures, if basic restoration measures cannot completely restore the damaged environment to the baseline state.*
> - *to select the best restoration plan based on restoration measures and implement it.*

14.1.2 In the European Union

The EU's system of compensation for environmental damage was established based on the lessons and experience from the American system of compensation for natural resources damage, but only limited to environmental damage resulting from occupational activities. Strict liability shall be applied to the damage caused by occupational activities that are hazardous or potentially hazardous to the environment or people's health, and fault liability shall be applied to the damage caused by other occupational activities (Bell, McGillivray, & Pedersen, 2013).

14.2 Necessity

There are three major purposes of founding a system of compensation for environmental damage, as summarized below.

(1) To establish the principle of assumption of responsibility for environmental damage

The principle of assumption of responsibility for environmental damage was stipulated in the EP Law to require those who are responsible for environmental damage to assume the liability for providing compensation and restoring the damaged environment, so as to resolve the dilemma that enterprise causes the pollution, while the public suffers from it and the government pays for it.

(2) To make up for the deficiency of the existing system

As regulated in the Constitution, the *Law of the People's Republic of China on Property Rights* (the Property Rights Law) and other relevant laws, on behalf of the State, the State Council shall exercise the proprietorship right of state-owned assets. However, in the current system, there are no provisions to regulate the specific subject to claim compensation, once natural resources, including mineral reserves, water resource, urban land, forests, mountains, grasslands, wasteland, mudflat, and so on, are damaged.

(3) To fulfill the government's statutory duty

To protect and improve people's production and living environment is the government's obligatory duty. Through the implementation of compensation for environmental damage, the damaged environment can be remedied and restored.

14.3 Related Policies

In December 2015, the *Pilot Scheme of the Reform of the Environmental Damage Compensation System* was issued by the General Office of the CCCPC and the State Council, requiring that the pilot scheme shall be deployed in selected provinces from 2015 to 2017 so as to establish the system of restoration and compensation for environmental damage, and to promote the construction of ecological civilization. In addition, the managerial and technical system of validation and assessment, the assurance of finance, and the operation mechanism of the environmental damage compensation system can be gradually formed through the pilot scheme in which the scope of compensation for environmental damage, the subject of responsibilities, the subject of claim, and solutions of damage compensation are determined. In 2018, this pilot scheme was implemented nationwide. By 2020, a preliminary environmental damage compensation system shall be constructed with clearly defined responsibilities, clear channels, standardized technologies, strong assurance, and effective compensation and policies in place.

Several methods were promulgated by the MEP, including the *Recommended Methods of Calculating the Amount of Damage Caused by Environmental Pollution (Version I)* in May 2011, the *Recommended Methods of the Validation and Assessment of Environmental Damage (Version II)* in October 2014, the *Recommended Methods of Environmental Damage Assessment during the Emergency Response to the Unforeseen Environmental Incidents* in December 2014, and the *Notice of Emergency Response to the National Emergent Environmental Incidents* was promulgated by the State Council in December 2014, to direct the implementation of the environmental damage compensation system.

14.3.1 Scope of Application

One shall be held liable for compensation for environmental damage under the following circumstances: ① emergent environmental incidents with relatively larger scale or above; ② the incidents of environmental pollu-tion or ecological destruction that occur in the key ecological function zones and the prohibited development zones designated in national or provincial main function zones planning; ③ other incidents with serious impact on the ecological environ-ment. Regarding the compensation for personal injury, loss of personal or collective property, and damage to the marine environment, the Tort Liability Law, the Marine Protection Law, and other related regulations shall be applied.

14.3.2 Scope of Compensation

The scope of compensation for environmental damage includes pollution cleanup cost, pollution remediation cost, the loss of ecological services during the period of remediation and restoration, the loss caused by permanent damage to ecological functions, and the costs of investigation, validation, and assessment of environmental damage and compensation ensued.

14.3.3 Claimant to Compensation

Empowered by the State Council, the provincial government, as the claimant to the compensation for environmental damage in local administrative jurisdiction, may assign relevant departments or agencies to work on the claim to compensation for environmental damage. Provincial governments shall institute the regulations of the starting conditions to file the claim of environmental damage, the evaluation procedures to select validation and assessment agencies, the demarcation of administrative jurisdiction, information disclosure, and other work. In addition, provincial governments shall clearly define the responsibilities in the claim of environmental damage by various government agencies, such as environmental protection, land resources, urban and rural housing construction, water conservancy, agriculture, forestry, etc. Moreover, the mechanism of supervising actions in the claim of compensation for environmental damage shall be established. If case of any breach of privilege, neglect of duty, favoritism, and irregularities during the claim of compensation for environmental damage, legal liability shall be pursued in accordance with relevant laws and regulations.

References

Bell, S., McGillivray, D., & Pedersen, O. W. (2013). *Environmental law*. Oxford: Oxford University Press.

Faure, M. G., & Liu, J. (2012). New models for the compensation of natural resources damage. *Kentucky Journal for Equine, Agriculture, and Natural Resources Law, 4*(2), Maastricht Faculty of Law Working Paper.

Faure, M. G., & Liu, J. (2014). Compensation for environmental damage in China: Theory and practice. *Environmental Damage Compensation, 31*, 240–321.

Kubasek, N. K., & Silverman, G. S. (2014). *Environmental law* (8th ed.). New Jersey: Pearson Education Inc.

Wilde, M. (2013). *Civil liability for environmental damage: A comparative analysis of law and policy in Europe and US* (2nd ed.). The Netherlands: Kluwer Law International.

Chapter 15
Cleaner Production

For many developing countries, industrialization has been adopted as one of the important alternatives for economic transform, and the crucial means and shortcut to achieve substantial economic growth. However, many severe adverse consequences, such as fast resources depletion, ecological deterioration and environmental pollution, have greatly offset the economic gains. As an effective measure to mitigate the conflicts between environmental protection and economic development, cleaner production (CP) has been widely recognized as one of the best pathways to achieve sustainable development (Zhang, 2000; Luken & Navratil, 2004). Considered being superior to end-of-pipe technologies for both environmental and economic reasons (Frondel, Horbach, & Rennings, 2007), CP is the integrated preventive environmental strategy for processes, products and services, whose origin could be traced back to the Principles and Creation of Non-Waste Technology and Production. In 1976, CP was put forward in the Proceedings of the International Seminar on Non-Waste Technology and Production organized by the Senior Advisors to Economic Commission for Europe (ECE) Governments on Environmental Problems. In 1979, clean technologies became an important policy of the Commission of the European Economic Community (CEEC). And, in 1985, three key criteria for clean technologies, less pollution discharged to the environment, less waste (low- and non-waste technologies), and less demand for natural resources (water, energy, and raw materials), namely, were identified by CEEC (Geiser, 2002). In addition, some demonstration projects for clean technologies, financed by the European Communities, were implemented in seven specific industrial sectors, including surface treatments, leather industry, textile industry, cellulose and paper industries, mining and quarrying, chemical industry, and agro-food. In 1992, CP was formed as the preventive, company-specific environmental protection initiative during the preparation of the Rio Summit which was a joint program of the United Nations Environmental Programme (UNEP) and United Nations Industrial Development organization (UNIDO). This program, adopted the concept from the Pollution Prevention Pays (3P) Program launched by the 3M Company in 1974, was aiming to reduce the environmental impact from industries and had acquired many supports from enterprises and countries, worldwide (Berkel, 2001). From the simple idea to produce with less waste, CP has developed into the concept to increase the resource efficiency of production, in general (Kaźmierczyk et al., 2011).

15.1 Initiation

In late 1980s, cleaner production was initially introduced into China to lead the shift of the environmental protection efforts from end-of-pipe to pollution prevention (Ortolano, Cushing, & Warren, 1999; Wang, 1999). In 1992, the first Cleaner Production Training Session was held in Xiamen (Fujian Province) through the collaborative support from UNEP and UNIDO. In 1993, CP strategy of China was proposed by the SEPA and the SETC (Cushing, Wise, & Hawes-Davis, 1999, Zhang, Yang, & Bi, 2013). In the same year, the research project "Promoting China's Cleaner Production" was commenced to cultivate the first batch of trainees in cleaner production audit (CPA), where the first CPA project for 29 companies was completed during this training session, under the supervision and financial support from UNEP and UNIDO. In order to promote CP development in China, ten cities (including Beijing, Shanghai, Tianjin, Chongqing, Shenyang, Taiyuan, Jinan, Kunming, Lanzhou and Fuyang) and five industries (including petrochemical, metallurgical, chemical, shipbuilding and light industries) were selected as the demonstration pilot cites and industrial sectors for CP implementation, as listed in *the Notice Concerning Demonstration Pilot Site Project for Cleaner Production Implementation* publicized by SETC in May 1999 (Fang & Côté, 2005).

15.2 Legal System Development

In general, the legal system of CP in China was established through four distinct periods, beginning stage, legalization stage, institutionalization stage, and perfection stage, respectively (Duan & Zhou, 2007; Zeng, Meng, Yin, Tam, & Sun, 2010; Qi, Hu, & Xiang, 2012).

15.2.1 The Beginning Stage (1973–1992)

In 1973, the idea of "prevention first, integration of prevention and governance" was put forward in *Some Regulations Concerning Environmental Protection and Improvement (on Trial Draft)*, promulgated by the State Council as one of the key pollution control policies, which was the first regulation concerning CP. In 1983, the concept of CP was further embodied in the *Some Regulations Concerning Technical*

Modification for Industrial Pollution Prevention, promulgated by the State Council. In 1989, the action plan of promoting CP was proposed by UNEP to introduce the principles and methodologies of CP into China. In 1992, CP became one of the important countermeasures for ecological degradation and environmental pollution, as explicitly specified in *the Ten Strategies on Environment and Development*, promulgated by the State Council.

During this stage, the importance of CP to environmental protection was clearly realized in the legislative process. However, the functions of CP had not been fully demonstrated, owing to the constraints from the irrationality of the existing industrial structure, in addition to lacks of adequate technical skills and sufficient capital support (Geiser, 2001).

15.2.2 The Legalization Stage (1993–2002)

In October 1993, promoting CP was listed as one of the key themes for the Second National Meeting on Industrial Pollution Prevention and Control Work. In March 1994, CP was specified as one of the important policies throughout *the China's Agenda 21* promulgated by the State Council. Since 1994, CP became the primary requirement clearly identified in amendments to many laws (including the Air Pollution Law, the Water Pollution Law, and the Marine Protection Law) and newly instituted laws (such as the Solid Waste Pollution Law, the Energy Conservation Law, and *the Ordinances of Management for Environmental Protection of Construction Project*). In addition, CP was specified as the important environmental protection measures in the 9th FYP (1996–2000) and *the Decisions on Several Issues Concerning Environmental Protection*, disseminated by the State Council. Furthermore, in April 1997, the guiding principles, basic framework and practical methods of CP were proposed to be integrated with existing environmental management system, as depicted in *the Some Opinions Concerning Promoting Cleaner Production*, promulgated by SEPA. Several provisions regarding CP, such as *the Notice Concerning Demonstration Pilot Site Project for Cleaner Production Implementation* (in May 1999) and *the Catalogue of Technical Guidance of Cleaner Production for Key Enterprises (the 1st Batch)* (in February 2000) were disseminated by SETC. In addition, three batches of *the Catalogue of Outdated Production Capacity, Technologies and Products to be Phased-Out* were also promulgated by SETC in January and December 1999, and June 2002, consecutively. Most importantly, the Cleaner Production Promotion Law was promulgated on June 29th, 2002, and became effective on January 1st, 2003.

During this stage, many laws, regulations and provisions concerning CP were promulgated by various government departments and agencies to facilitate the legislation, legalization, institutionalization and standardization of CP, and to promote the researches and exploration on CP implementation. Additionally, several demonstration projects, pilot site studies and international cooperation projects pertaining to CP were implemented to speed up CP development in China, and to catch up with modern trend and most advanced CP technologies, worldwide.

15.2.3 The Institutionalization Stage (2003–2005)

The proclamation and effectiveness of the PCP Law marked the milestone of CP development in China. Several corresponding rules, provisions standards, and management schemes were publicized to realize this special law. In 2003, *the Catalogue of Technical Guidance of Cleaner Production for Key Enterprises (the 2nd Batch)* was disseminated by SEPA and SETC, jointly (in February); *the Some Opinions Concerning Fully Implementing the Law of People's Republic of China on Promoting Cleaner Production* was disseminated by SEPA (in April); and *the Some Opinions Concerning Expediting The Implementation of Cleaner Production* was disseminated by the State Council (in December). In 2004, *the Interim Measures of Cleaner Production Audit* was promulgated by SEPA (in August), in which voluntary cleaner production audit (VCPA) and mandatory cleaner production audit (MCPA) were clearly defined; and *the Measures of Utilizations and Management for the Cleaner Production Special Funds Subsidized from Central to Local Governments* was promulgated by the MOF (in October) to regulate the utilizations and management for the Cleaner Production Special Funds. In December 2005, *the Regulations on the Procedures of Cleaner Production Audit for Key Enterprises* was promulgated by SEPA, along with *the Catalogue of Toxic and Hazardous Materials Requiring Cleaner Production Audit (the 1st Batch)*, where several critical requirements for CPA were explicitly specified, including the criteria for screening key enterprises, the information for public announcement, the qualifications of consulting agencies, CPA procedures and acceptance criteria, and so on. Moreover, the Cleaner Production Assessment Indicator Systems for various industrial sectors were promulgated consecutively by the NDRC and SEPA, since 2005.

During this stage, various management measures related to CPA, such as voluntary and mandatory, procedures, criteria, requirements, normative documents, administration and supervision, and assessment and acceptance, were clearly standardized to greatly facilitate CP implementation in China. Furthermore, many industries have acquired significant practical experience for future improvement, through active participation in CP exercises.

15.2.4 The Perfection Stage (2006 to Date)

In November 2006, *the Catalogue of Technical Guidance of Cleaner Production for Key Enterprises (the 3rd Batch)* was disseminated by SEPA and SETC, jointly. In July 2008, *the Notice Concerning Further Enhancement on Cleaner Production Audit for Key Enterprises* was promulgated by MEP, along with *the Implementation Guidance* on Assessment and Acceptance of Cleaner Production Audit for Key Enterprises (on Trial) and *the Catalogue of Toxic and Hazardous Materials Requiring Cleaner Production Audit (the 2nd Batch)*, to regulate CPA, to encourage and guide companies to carry out CP effectively, and to ensure the effectiveness of energy conservation and emission reduction. In September 2009, *the Notice Concerning Enhancement on Promoting Cleaner Production for Industries and Information Technology Sectors* was promulgated by the MIIT to encourage energy conservation, emission reduction, structure modification and transformation of economic development modes for industrial and information technology sectors. In October 2009, *the Interim Measures of Management for the Central Finance Cleaner Production Special Funds* was promulgated by MOF and MIIT, jointly, to regulate the management for the Central Finance Cleaner Production Special Funds, to increase the benefits of funds utilizations, to speed up the promotion and applications of CP technologies, and to promote CP implementation in various industrial sectors. In 2010, *the Notice Concerning Further Promoting Cleaner Production for Key Enterprises* was promulgated by the MEP, along with *the Catalogue of Business Classification Management for Cleaner Production for Key Enterprises*, to explicitly categorize 21 different major industrial sectors. And, as suggested in the Amendment of the Cleaner Production Promotion Law proclaimed on February 29th, 2012, the planning system of CP implementation and CPA system should be established to further accelerate CP implementation in China more effectively and efficiently.

During this stage, the institution of the mechanisms for CPA assessment and acceptance, an important component of China's CP policies, has played the essential role in the perfection and innovation for CPA system, which is very significant to the effectiveness of CP programs and the quality assurance of CPA for industries.

All these important laws, regulations, provisions and official documents concerning promoting CP development at four different stages were summarized in Table 15.1.

15.3 Capacity Building

15.3.1 Institution Development

In December 1994, the National Cleaner Production Center (NCPC) was founded under the approval of SEPA, sponsored by UNEP/UNDIO, in order to provide a quasi-standardized model for national CP service delivery in developing and transition economies. Indeed, NCPC has been playing the crucial role in promoting CP development for the past two decades (Khalilia, Duecker, Ashton, & Chavez, 2015). Ever since, many CP centers for various industrial sectors and CP consulting service institutions were established, nationwide. By 2002, there were only 39 CP centers and consulting service institutions, including 10 CP centers for various industrial sectors, 21 CP consulting service institutions at provincial level, and 8 CP consulting

Table 15.1 List of official documents concerning cleaner production at different stages

Period	Year	Government Agency	Laws, Regulations, Provisions, and Official Documents
Stage I	1973	State Council	Some Regulations Concerning Environmental Protection and Improvement (on Trial Draft)
	1983	State Council	Some Regulations Concerning Technical Modification for Industrial Pollution Prevention
	1992	State Council	Ten Strategies on Environment and Development
Stage II	1993	SEPA and SETC	The Second National Meeting on Industrial Pollution Prevention and Control Work
	1994	State Council	China's Agenda 21
	1996	State Council	The 9th FYP (1996–2000)
	1996	State Council	Decisions on Several Issues Concerning Environmental Protection
	1997	SEPA	Some Opinions Concerning Promoting Cleaner Production
	1998	State Council	The Ordinances of Management for Environmental Protection of Construction Project
	1999	SETC	Catalogue of Outdated Production Capacity, Technologies and Products to be Phased-Out (the 1st Batch)
	1999	SETC	Notice Concerning Demonstration Pilot Site Project for Cleaner Production Implementation
	1999	SETC	Catalogue of Outdated Production Capacity, Technologies and Products to be Phased-Out (the 2nd Batch)
	2000	SETC	Catalogue of Technical Guidance of Cleaner Production for Key Industries (the 1st Batch)
	2002	SETC	Catalogue of Outdated Production Capacity, Technologies and Products to be Phased-Out (the 3rd Batch)
	2002	State Council	The Law of People's Republic of China on Promoting Cleaner Production
Stage III	2003	SEPA & SETC	Catalogue of Technical Guidance of Cleaner Production for Key Industries (the 2nd Batch)
	2003	SEPA	Some Opinions Concerning Fully Implementing the Law of People's Republic of China on Promoting Cleaner Production
	2003	State Council	Some Opinions Concerning Expediting The Implementation of Cleaner Production
	2004	SEPA	Interim Measures of Cleaner Production Audit
	2004	MOF	Measures of Utilizations and Management for the Cleaner Production Special Funds Subsidized from Central to Local Governments

(continued)

15.3 Capacity Building

Table 15.1 (continued)

Period	Year	Government Agency	Laws, Regulations, Provisions, and Official Documents
	2005	SEPA	Regulations on the Procedures of Cleaner Production Audit for Key Enterprises
	2005	SEPA	Catalogue of Toxic and Hazardous Materials Requiring Cleaner Production Audit (the 1st Batch)
	2005	NDRC & SEPA	The Cleaner Production Assessment Indicator System for various industries
Stage IV	2006	SEPA & SETC	Catalogue of Technical Guidance of Cleaner Production for Key Industries (the 3rd Batch)
	2008	MEP	Notice Concerning Further Enhancement on Cleaner Production Audit for Key Enterprises
	2008	MEP	Implementation Guidance on Assessment and Acceptance of Cleaner Production Audit for Key Enterprises (on Trial)
	2008	MEP	Catalogue of Toxic and Hazardous Materials Requiring Cleaner Production Audit (the 2nd Batch)
	2009	MIIT	Notice Concerning Enhancement on Promoting Cleaner Production for Industries and Information Technology Sectors
	2009	MOF & MIIT	Interim Measures of Management for the Central Finance Cleaner Production Special Funds
	2010	MEP	Notice Concerning Further Promoting Cleaner Production for Key Enterprises
	2010	MEP	Catalogue of Business Classification Management for Cleaner Production for Key Enterprises
	2012	State Council	The Amendment of the Cleaner Production Promotion Law
	2013	NDRC, MEP, MIIT	General Provisions of Compiling the Cleaner Production Assessment Indicator System (on Trial)
	2014	MIIT	Implementation Procedures of Cleaner Production Technologies for Key Enterprises and Industries in Air Pollution Prevention and Control
	2015	MIIT	The Standards of Cleaner Production Audit for Industries and the Standards of Effectiveness Evaluation of Cleaner Production Audit for Industries
	2016	NDRC & MEP	Measures of Cleaner Production Audit

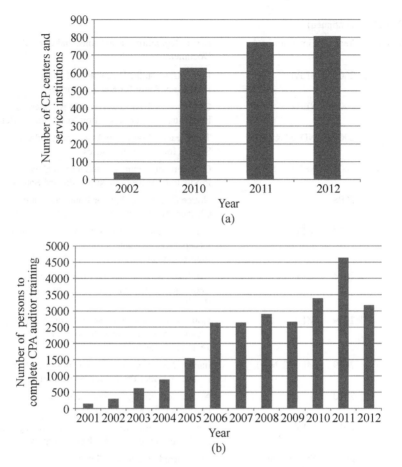

Fig. 15.1 Number of CP centers and CP service institutions, and persons to complete CPA auditor training

service institutions at city level. During this 8-year period, the fundamental construction was slow. However, the development was rapid and vigorous for the next 8-year period. By 2010, there were 629 CP centers and consulting service institutions. And by 2012, the number reached 806, as shown in Fig. 15.1a. These CP centers and consulting service institutions had developed into a nationwide CP service network to provide strategic and technical services for various industrial sectors.

15.3.2 Professional Training

From 2001 to 2012, more than 400 "National Cleaner Production Auditors Training Sessions" were held by national training institutions. During these training sessions, more than 25,000 persons were educated, trained and qualified as CP auditors, as shown in Fig. 15.1b. In addition, many CP training sessions were held by local training agencies to cultivate more CP practitioners to be more familiar with CP policies, technologies and skills, through demonstration projects and hands-on experience.

15.3.3 Guidelines and Standards

By the end of 2013, there were 58 different Industrial Standards of Cleaner Production for various industrial sectors, 2 batches of the Catalogue of Toxic and Hazardous Materials Requiring Cleaner Production Audit, and Cleaner Production Assessment Indicator Systems for 45 different industries, promulgated by NDRC and MEP, consecutively. In addition, 3 batches of the Catalogue of Technical Guidance of Cleaner Production for Key Enterprises were promulgated by MIIT. These guidelines and standards provided concrete foundation and technical supports in promoting and facilitating CP development.

15.4 Mandatory Cleaner Production Audit

Since the promulgation of *the Interim Measures of Cleaner Production Audit*, CPA has become the major scheme to promote CP, in which MCPA was implemented by diverse key enterprises, nationwide, and the lists of these companies were publicized in batches, since 2007. By 2012, there were about 27,778 companies to implement MCPA, as shown in Fig. 15.2a.

In encouraging, recognizing and awarding companies for their participation and accomplishments in CP, and successful completion of CPA, *the Announcement of Key Enterprises to Pass Cleaner Production Audit (the 1st Batch)* was disseminated by MEP on September 6th, 2010 to clearly list the companies who have implemented CP and successfully past CPA assessment and acceptance. *The 2nd Batch of Announcement* (December 8th, 2010), *the 3rd Batch of Announcement* (July 1st, 2011), *the 4th Batch of Announcement* (December 31st, 2011), and *the 5th Batch of Announcement* (September 12th, 2012) were then publicized, successively. A total amount of 17,863 companies were listed in these five batches of announcement, as shown in Fig. 15.2b. As noticed, the number in *the Fifth Batch* increased considerably to reach 8,775, which is more than 49% of the total amount of the 5 batches, to indicate that CP development in China has gradually reached the booming phase.

As stated in the Amendment of the Cleaner Production Promotion Law (2012), the MCPA became the mandatory obligation for enterprises, in which firms are required to undertake MCPA under the following three conditions: ① discharging pollutants beyond the national or local emission standards or, though attainment to the national

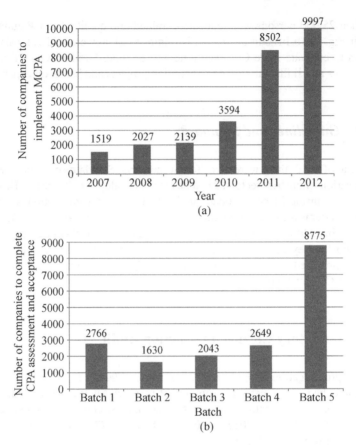

Fig. 15.2 Number of companies to implement MCPA and complete CPA assessment and acceptance

or local emission standards, but exceeding the emission cap of major pollutants; ② exceeding the limit of energy consumption per unit of products to become high energy consumption; ③ using toxic or harmful raw materials for production or discharging toxic or harmful substances during production. The detailed procedures of MCPA are listed in Box 15.1.

Box 15.1 Detailed procedures of MCPA

- *The competent authorities at city level propose a list of firms required to undertake MCPA within their administrative jurisdiction and submit the list to the competent authorities at provincial level for approval.*
- *The competent authorities at provincial level approve the list and disclose the information on internet site.*

> - *The firms on the list should disclose information on production, pollution discharge, energy consumption, etc. on designated internet site within one month.*
> - *The firms on the list should accomplish the MCPA within one year. The firms could undertake the MCPA either by themselves or commission a consultancy. The MCPA report should be submitted to the competent authorities at city level.*
> - *The competent authorities at city level is responsible to review the MCPA report and make a decision on whether the MCPA report is acceptance.*

15.5 Recommendations for Promoting CP Development in China

In order to further facilitate and promote CP development in China, many actions and measures need to be taken.

(1) To perfect the institutional system

In order to resolve the discontinuity and inconsistence between diverse laws and regulations concerning CP, it is necessary to institute supporting and supplemental regulations and policies, to enhance the pertinence, compulsion, and enforceability of laws and provisions, and to increase the applicability of CPA to various industries. In addition, it's essential to expand CP to the primary industries and tertiary industries, which can be greatly improved and benefited from CP. Moreover, it's highly recommended to integrate CP with other environmental management schemes, for example, emission reduction and energy conservation, total emission control, environmental impact assessment and emission permit, to ensure the effectiveness of CP.

(2) To improve administrative management

As clearly stipulated in the PCP Law, the administrative jurisdiction of MPCA is the competent authority of environmental protection. However, it's necessary to refine the duty and scope of work for various government departments and agencies, and to establish the collaborative procedures and coordinating mechanisms between different government departments and agencies, such as environmental protection bureau, the competent authority of industrial sector, and the funding management agency. Also, it's important to establish the management system for CP consulting institution and CP practitioners to regularly supervise the quantity and quality of CP implementation. Furthermore, it is essential to institute the charge and fee standards for CP consulting services to regulate a normal and competitive market of CP consulting.

The CP consulting institutions should be classified into various levels according to their specialty, professional staff, technical capacity, business volume, and so on.

(3) To expand funding sources

For both central and local governments, it is necessary to institute some promoting and preferential policies, including emission reduction incentives, cleaner production subsidy, tax exemption and reduction, and bank financing, to encourage and reward enterprises to participate in VCPA. In addition to the funding support from the Central Government, local governments should set up a budget plan for MCPA, where partial funds can be raised through organizing CP training sessions for practitioners, as the regular on-job training sessions. Moreover, for significant achievement in emission reduction, energy conservation and resources conservation through CP implementation, it is feasible to provide special financial support, from the Environmental Protection Special Funds, the Circular Economy Special Funds, the Heavy Metal Pollution Prevention Funds, and/or Returning Funds from Pollution Fee, to further encourage CP implementation.

(4) To upgrade technical capacity

In order to enhance the professional capability, it's urgent to establish the continuous supervising and managing mechanisms, including regular on-job training sessions, annual performance evaluation and professional qualification certificate system, for CP administrative staff and practitioners of CP consulting institutions. The aims of these mechanisms are to update the most recent laws, regulations and standards, and the newest technologies, to exchange experience learned from practices and case analyses, to assess the performance of CP practitioners, and to supervise the market behavior of CP consulting institutions. It's also advised to establish the mechanism to encourage CP consulting institutions and individual CP practitioner to regularly compile and publish reports and books, for example, technical summary and case analyses, to facilitate the dissemination of knowledge and technologies.

(5) To enhance acceptance system

Through the following steps, CPA assessment and acceptance system can be greatly promoted. First, it is necessary to establish a national database of experts for CPA assessment and acceptance for key enterprises. The qualifications of the experts to be listed in this database should be determined by MEP. These experts with diverse expertise can be recruited by MEP or through competitive applications. As the think tanks, the main task of these experts is to provide technical and intelligent support, and to participate in CPA assessment and acceptance for key enterprises. These experts are regularly monitored and evaluated for their performance, and their terms should be limited and conditioned. Second, in order to facilitate the execution of CPA assessment and acceptance, it is crucial to constantly promulgate updated technical guidelines for various industrial sectors to direct CP implementation, and to institute the standard system for CPA assessment and acceptance to quantitatively evaluate CP achievement. Third, it's advised to establish the post-assessment mechanisms to ensure full and continuous CP implementation as recommended and approved.

(6) To increase research projects

First, it's necessary to increase the funding support toward CP research projects, including regulations, policy, institution, methodologies, technical guidelines, evaluation and demonstration projects, in order to build a more concrete foundation for CP development, and to explore more aspects of CP implementation. Second, in order to promote the public participation, it's essential to develop a platform of information disclosure to timely broadcast new laws, regulation, provisions and latest development, to publicize the necessary information of consulting institution, to disseminate important news, and to exchange experience learned from practice.

15.6 Conclusions

Since CP was firstly introduced into China in late 1980s, the legal system and technical framework of CP have been gradually improved and perfected through numerous pilot studies, demonstration projects and practices, all over the country, for more than 30-year development. After the first 10 years of exploring and learning, the Cleaner Production Promotion Law was promulgated in 2002 to facilitate CP development. For the next 10 years of practice and implementation, the Amendment of the Cleaner Production Promotion Law was promulgated in 2012 to further promote CP advancement. There are many remarkable achievements, through 20 years of CP implement. In addition, though CP development in China is generally on upward trend, noticeably, various regional approaches for CP implementation have evolved, due to significant geographical differences between various regions, and considerable local peculiarities and disparity on industrial allocation, especially for key enterprises, due to governmental initiatives (Hicks & Dietmar, 2007). Hence, it would be great beneficial to CP development in China if the regionalized CP management system is established and operated, based on essential geographical characteristics, diverse regional differences and various industries allocation.

References

Berkel, R. V. (2001). Cleaner production perspectives 1: CP and industrial development. *UNCP Industry and Environment, 24*(1–2), 28–32.
Cushing, K. K., Wise, P. L., & Hawes-Davis, J. (1999). Evaluating the implementation of cleaner production demonstration projects. *Environmental Impact Assessment Review, 19*, 569–586.
Duan, N., & Zhou, C. (2007). Research and analysis of the formation process of compulsory regulation and policy of cleaner production audit. *China Population, Resources and Environment, 17*(4), 107–110.
Fang, Y., & Côté, R. P. (2005). Towards sustainability: Objectives, strategies and barriers for cleaner production in China. *International Journal of Sustainable Development and World Ecology, 12*(4), 443–460.

Frondel, M., Horbach, J., & Rennings, K. (2007). End-of-pipe or cleaner production? An empirical comparison of environmental innovation decisions across OECD countries. *Business Strategy and the Environment, 16,* 571–584.

Geiser, K. (2001). Cleaner production perspectives 2: Integrating CP into sustainability strategies. *UNEP Industry and Environment, 24*(1–2), 33–36.

Geiser, K. (2002). What next in cleaner production technologies? *UNEP Industry and Environment, 25*(3–4), 75–77.

Hicks, C., & Dietmar, R. (2007). Improving cleaner production through the application of environmental management tools in China. *Journal of Cleaner Production, 15,* 395–408.

Kaźmierczyk, P., Hansen, M. S., Günther, J., McKinnon, D., Loewe, C., Lingvall, F., et al. (2011). Resource efficiency in Europe. *Copenhagen, European Environment Agency, 80.*

Khalilia, N. R., Duecker, S., Ashton, W., & Chavez, F. (2015). From cleaner production to sustainable development: The role of academia. *Journal of Cleaner Production, 96,* 30–43.

Luken, R. A., & Navratil, J. (2004). A programmatic review of UNIDO/UNEP national cleaner production centres. *Journal of Cleaner Production, 12,* 195–205.

Ortolano, L., Cushing, K. K., & Warren, K. A. (1999). Cleaner production in China. *Environmental Impact Assessment Review, 19,* 431–436.

Qi, Y., Hu, T., & Xiang, Z. (2012). Decennial review on cleaner production and its prospect in the twelfth five-year-plan (I): Decennial review on cleaner production in China. *Technology and Economics in Petrochemicals, 28*(1), 11–16.

Wang, J. (1999). China's national cleaner production strategy. *Environmental Impact Assessment Review, 19,* 437–456.

Zeng, S. X., Meng, X. H., Yin, H. T., Tam, C. M., & Sun, L. (2010). Impact of cleaner production on business performance. *Journal of Cleaner Production, 18,* 975–983.

Zhang, T. (2000). Policy mechanisms to promote cleaner production in China. *Journal of Environmental Science and Health, Part A: Toxic/Hazardous Substances & Environmental Engineering, A35*(10), 1989–1994.

Zhang, B., Yang, S., & Bi, J. (2013). Enterprises' willingness to adopt/develop cleaner production technologies: An empirical study in Changshu, China. *Journal of Cleaner Production, 40,* 62–70.

Chapter 16
Ecological Compensation

16.1 Introduction

Since the early 1970s, governments in developed countries have initiated plans to transform resource-based industries into eco-friendly and less resource consumptive industries to support sustainable development. During the implementation of economic transformation plans, government leadership and high compensation payments for economic transformation have become two essential tools for ensuring the effectiveness and efficiency of economic transformation. Ecological compensation can be interpreted as the payments for ecosystem services and natural resource protection, compensation for the opportunity cost of development, payments for ecosystem damage and natural resource consumption, and the cost of environmental pollution, which is paid through monetary or economic compensation. Therefore, the ecological compensation mechanism (ECM) is an important economic means to impose restrictions on environmentally damaging development, provide economic incentives for eco-friendly development, encourage environmental protection, and establish sustainable and secure financial funds for ecological restoration and environmental remediation.

Through government and market techniques, the ECM is one of the most important tools that regulate stakeholder relationships involving environmental protection and the extraction and utilization of natural resources. This technique is based on the value of ecosystem services, the cost of environmental protection, and the opportunity cost of development. In general, an ECM is established according to the available natural resources, population, economic structure, and environmental capacity to institute diverse environmental protection goals for ecologically fragile areas. There are three different levels of ECM implementation: macro-level, mid-level and micro-level. At the macro-level, the central government is the main body that implements ECM programs for individuals who benefit from ecosystem services or those responsible for ecosystem destruction. At the mid-level, regional ECM programs, such as ECMs for watersheds and ECMs between local governments, are implemented by local governments. At the micro-level, a transaction compensation program is implemented, such as compensation between beneficiaries of the exploitation and utilization of natural resources and local residents who incur the loss of ecological benefits and the opportunity cost of development.

In China, due to contrasting economic conditions between urbanized areas and the places of origin of natural resources, the major policies for regional development strategies are quite diverse, resulting in an unbalanced and irrational approach to regional sustainable development. While many ECM pilot cases were implemented by central and local governments, limited achievements have been documented due to major issues and obstacles concerning ECMs for natural resource use at various levels, such as policies and institutions, administration and management, and finance and capital allocation. The main content of this chapter covers the current status and existing problems related to ECMs for natural resource utilization in China. A variety of cases at different levels (macro-level, mid-level and micro-level) are analyzed to provide insights and practical recommendations for the institution and implementation of ECMs. Finally, several suggestions for future research are also illustrated.

16.2 Development of ECMs

In developed countries, ECM programs are established based on advanced public finance systems and are implemented through subsidy policies and financial transfer payments. In the United States, there are three types of ECM programs codified under the *Agri-Environmental Policy*: voluntary, incentive-based programs, regulatory programs, and cross-compliance programs, which are generally focused on compensation for soil and water conservation. The compensation is calculated based on the opportunity cost of agricultural crops. In the EU, according to the *Common Agricultural Policy*, ECM programs, including direct subsidy and transfer payments, are voluntary regulatory mechanisms focused on compensation related to agricultural protection, such as natural landscapes and cultural heritage, and the practices of environmental-friendly and low-intensity agriculture, such as the development of organic agriculture. The compensation is calculated based on the opportunity costs to local farmers (Baylis, Peplow, Rausser, & Simon, 2008). In Britain, the *North York Moors Farm Scheme*, a policy of ecological protection compensation, was launched in April 1990. In this scheme, compensation payments were offered for the sensitive management of important wildlife habitats, such as flower-rich grasslands and native woodlands, and the maintenance of walls and hedges. In Switzerland, the ecological protection compensation system is embodied in the *Federal Law on Agriculture*, enacted on April 29, 1998, to provide compensation for eco-friendly practices. In Australia, irrigators provide compensation for upstream reforestation and vegetation.

In developing countries, ECM projects are generally financed by developed countries and international funds, based on a balance between environmental protection and economic development. For example, the Integrated Conservation and Development Project (ICDP) has become the main body of the ECM program for many

16.2 Development of ECMs

developing countries in Asia and Africa (Rondeau & Bulte, 2007). The World Bank sponsored the establishment of the ECM framework for watershed services in Latin American nations, including Costa Rica, Ecuador, Colombia, Mexico, Brazil, etc. Costa Rica began implementing ECM schemes in early 1995 and became the global pioneer for the environmental services payment scheme (Pagiola, Arcenas, & Platais, 2005). However, very few concrete achievements have been reported due to severe conflicts between environmental protection and the demands of economic development. For example, economic growth often results in the demand for a higher quality of life, which amplifies the exploitation and utilization of natural resources. In addition, the economic schemes for ecological compensation may generate legal justifications for the excessive exploitation and utilization of natural resources that cause unnecessary environmental damage and hasten natural resource depletion.

In China, a planned economy and strong demands for growth in the gross domestic product (GDP) have caused excessive exploitation and inefficient utilization of natural resources, which have not only caused rapid natural resource depletion but also resulted in severe conflicts between environmental, economic and social benefits (Chang & Wu, 2011). Although ECMs advocate and implement resource preservation, ecological conservation and environmental protection, several inevitable consequences have emerged, including an unbalanced economic structure and social inequity. The emissions charges system was instituted in China in 2002 and was regarded as the only environmental management scheme related to ECMs. Unfortunately, it is not truly an ECM but a mechanism by which the competent authorities for environmental protection collect fees of pollution prevention and control, and administrative management. Therefore, there is an urgent need to establish ECMs. As explicitly indicated in *the Decision on Implementing the Scientific Outlook on Development and Strengthening Environmental Protection*, promulgated in December 2005, and the *11th FYP (2006–2010)*, promulgated in 2006, it is necessary to establish ECMs to minimize the unequal distribution of benefits and mitigate conflicts of interest between different stakeholders as soon as possible, particularly for the exploitation and utilization of natural resources.

In China, there are abundant and diverse natural resources. However, the average available resource per capita is quite small in comparison with the world average because of the enormous population (Peng, 2011), as shown in Table 16.1. Therefore, the effective and efficient exploitation and utilization of natural resources has become the basis for economic and social development strategies. Additionally, it is necessary to establish appropriate and reasonable ECMs to protect the ecosystem, ensure sustainable development, regulate the interests between stakeholders, and improve environmental equity.

In the early 1990s, the practice of ECM in China was first initiated solely for forest ecological benefit compensation. To facilitate ECM implementation, the *Ordinance of Ecological Compensation* was drafted in April 2010 and covered several key fields, including forest, grassland, wetland, resource exploitation, marine, watershed and ecological functional zones. Currently, this *Ordinance* has not been officially promulgated. Yet, there are presently four major areas of ECM implementation, such as watershed, resource exploitation, ecosystem services and nature reserves where the

Table 16.1 Total amount and availability of natural resources in China

Natural resources		Total	Unit	Per capita	Unit
Population		1.341	10^9 capita		
Land		9.600	10^6 km^2	0.00716	km^2
	Cultivated land	1.217	10^6 km^2	0.00091	km^2
Forest[a]		1.950	10^6 km^2	0.00145	km^2
	Standing trees and logs[b]	13.721	10^9 m^3	10.234	m^3
Grassland		4.000	10^6 km^2	0.00298	km^2
Fresh water		2,812.400	10^9 m^3	2.098	10^3 m^3
Primary energy	Coal	114.500	10^9 ton	0.085	10^3 ton
	Natural gas	1,900.000	10^9 m^3	1.417	10^3 m^3
	Petroleum	2.120	10^9 ton	1.581	10^3 ton
	Hydropower	1,760.000	10^9 kW·h	1.313	10^3 kW·h
Primary energy	Solar energy	1,700.000	10^9 TCE[c]	1.268	10^3 TCE
	Wind energy	1.000	10^9 kW	0.746	kW
	Thermal energy	3.160	10^9 TCE	2.357	TCE
	Biomass energy	0.260	10^9 TCE	0.194	TCE

[a]Forest: Forest functions as habitat for organisms, hydrologic flow modulator, soil conserver, and so on, measured in area (km^2)
[b]Standing tree and logs: is one of the ecosystem services provided by forest, measured in volume (m^3)
[c]TCE: tons of coal equivalent

major recipients of compensation are regional administrations, local governments, or farmers and herders, as shown in Table 16.2.

16.3 Financial Sources for ECM in China

Currently in China, the central government plays a leading and vital role in the establishment of ECM programs through financial transfer payments, special funds, resource taxation systems, regional policies, major ecological protection engineering projects, etc.

16.3.1 Financial Transfer Payment

Financial transfer payment is the financial policy that promotes balance and coordination between economic and social development in various regions and is one of the

16.3 Financial Sources for ECM in China

Table 16.2 Major areas of ECM implementation in China

Area	Major ECM approaches	Means of ECM
Watershed	Compensations between upstream and downstream for large-scale watershed Compensations between trans-provinces for mid-scale watershed Compensations between regional administrations for small-scale watershed	Negotiation between local governments and regional administrations Financial transfer payment Market trade
Resources exploitation	Compensations for land reclamation Compensations for vegetation restoration	Payment by beneficiary Payment by destroyer Payment by developer
Ecosystem services	Compensations for forest ecosystem services	Financial transfer payment by national public compensation Ecological compensation funds Market trade Involvement of enterprises and individuals
Nature reserves	Water source conservation areas Biodiversity reserves Wind-breaking and sand-fixing areas Soil conservation areas Flood regulating reservoirs	Compensations from central and local governments Donations received from NGOs Involvement of enterprises and individuals

major regional compensation mechanisms for transferring national revenue to local social welfare and financial subsidy programs. Thus, it is an effective means to mitigate vertically (central to local) and laterally (local to local) unbalanced economic conditions between central and local governments. In China, to minimize the huge differences in regional economic development, the financial transfer payment has become the major tool for ECM since 1994. Through financial transfer payments, the central government has invested more than ¥4.00×10^{10} in ecological engineering projects, such as returning farmland to forest/grassland, natural forest protection, anti-desertification, Beijing-Tianjin sandstorm sources governance, returning rangeland to grassland, and large-scale shelter forest plantations in northwest, north and northeast China (the "Three-North Project"). Some local governments have also instituted incentive programs, including financial subsidies, environmental protection and restoration subsidies, public benefits for eco-forest subsidies and ecological construction objectives awards, to encourage all ecological conservation, environmental protection, and related activities. Moreover, other local governments have adopted financial programs as important measures to support environmental protection actions in mountain areas.

16.3.2 Special Funds

To effectively and efficiently implement ECMs, special funds are an important financial source for ECM programs that are initiated and managed by government departments and agencies, including the MLR, the SFA, the MWR, the MOA and the MEP. To provide capital subsidies and technical support for environmental protection and construction, various policies were instituted and special funds were established, including the New Energy Construction for Villages, Soil and Water Conservation Subsidy, Farmland Protection and Public Benefits Eco-Forest Subsidy. In 1999, the Biogas Engineering Projects Subsidy was founded by the MOA and the Forestry Ecological Benefit Compensation Funds were established by the SFA. The Small-Scale Farmland Water Resources Subsidy and Soil and Water Conservation Subsidy were jointly established by the MWR and the MOF. In addition, the Public Benefits Eco-Forest Subsidy in Guangdong and Zhejiang provinces provide ¥12,000/km^2 and ¥10,500/km^2, respectively.

16.3.3 Resource Taxation System

Usually, resource taxes are paid to provincial governments when the resources are extracted, before they are refined or processed into other products. This taxation has become an important capital source for financially strapped local governments. On September 18, 1984, the resource tax system was first implemented through the promulgation of *the Ordinances for the Resources Taxes (Draft)*, which was replaced by *the Interim Ordinances of the Resources Taxes* (the Resource Tax Ordinances), effective January 1st, 1994. The Resources Taxes Ordinances were amended, promulgated on September 21st and became effective on November 1st, 2011. The Resource Tax Ordinances, as an attempt to reduce the fast-growing nation's dependency on resources, is an important step in China's progress toward a cleaner, more efficient growth model that is less dependent on resource-intensive industries.

Prior to the Resource Tax Ordinances, resource tax was calculated based on the volume of production, as opposed to the sales value. The new resource taxation system will shift profits from companies to governments in poorer provinces, which is an important way for China to boost revenues for provincial governments. Within the Resource Tax Ordinances, seven different categories of natural resources are listed: crude oil, natural gas, coal, non-metal mining, ferrous metal mining, non-ferrous metal mining, and salt. In addition, there are two types of resource tax rates in the Resource Tax Ordinances: sales value-based and sales volume-based tax rates. The sales value-based tax is applied to crude oil and natural gas, while the sales volume-based tax is applied to the other five categories, as shown in Table 16.3.

Table 16.3 Major resource taxes listed in the resource tax ordinances

Item		Tax rate	Tax type
1. Crude oil		5–10% of sales value	Sales value-based tax
2. Natural gas		5–10% of sales value	
3. Coal	Coking coal	¥(8–20)/ton	Sales volume-based tax
	Others	¥(0.3–5)/ton	
4. Non-metal	Ordinary	¥(0.5–20)/ton or m^3	
	Precious	¥(0.5–20)/kg or carat	
5. Ferrous metal		¥(2–30)/ton	
6. Non-ferrous metal	Rare-earth ores	¥(0.4–60)/ton	
	Other non-ferrous metal	¥(0.4–30)/ton	
7. Salt	Solid	¥(10–60)/ton	
	Liquid	¥(2–10)/ton	

16.3.4 Payments for Ecological Compensation

There are three different methods of payment for ecological compensation funds: central to local, intra-province and inter-provinces.

(1) Central to local

Financial transfer payments are based on unbalanced local financial budgets, particularly for remote and rural areas and autonomous regions with ethnic minorities. The payment is equal to the local financial revenue minus the average national financial revenue minus taxation. Special funds are the major capital sources for ECMs and are administered by the MOF, the NDRC, and other related departments, to offer compensation to ecological service providers. For instance, in "Returning Farmland to Forest", grain compensation is ¥0.35/kg for 1.50×10^5 kg/(km^2 · yr), the one-time seeding fee for forest plantations is $¥7.5 \times 10^4$/km^2, and the living subsidy is $¥3.0 \times 10^4$/(km^2 · yr). In "Grassland Enclosure", the one-time fencing construction fee is $¥3.0 \times 10^4$/km^2, of which the central government provides 70% of the fee as a subsidy and the rest is raised by local governments and individuals. The fodder subsidy is ¥22.5/(kg · day · km^2) from the central government.

(2) Intra-province

This is the "vertical" financial transfer payment from provincial governments to county governments, which can be paid either in cash (subsidies for grain allocation, transportation, and labor and operation costs) or through projects (integrated agricultural development, antipoverty development, and water and soil conservation). Additionally, to promote and facilitate local economic development, provincial governments create opportunities and conditions for county governments to improve local conditions, such as infrastructure construction, economic development (science, technology, culture, education and business), and social welfare (higher social security and fewer employment problems).

(3) Inter-province

This is the "lateral" financial transfer payment between provinces within the same watershed, especially for the Yangtze River, Yellow River, etc. There are two different means of payment from provinces in the middle and lower streams to provinces in the upper streams: direct payment (cash, food and grain, and other resources) and indirect payment (business investment and cooperation, acceptance and allocation of ecological immigrants from upper streams, and other forms of one-to-one assistance).

16.4 Major Ecological Engineering Projects in China

Through ecological engineering projects, environmental quality has been greatly improved and various compensation packages, such as capital, resources, and technology, have been provided to residents living in project areas. As mentioned in Sect. 3.1, currently, there are several ongoing, large-scale ecological engineering projects, including the Three-North Project, natural forest protection, the Beijing-Tianjin sandstorm sources governance, returning farmland to forest/grassland and returning rangeland to forest/grassland.

16.4.1 Three-North Project

The Three-North Project, one of the largest ecological engineering projects in the world, with an investment of ¥5.77×10^{10}, was initiated in November 1978 and will continue until 2050. This project focuses on ecological restoration and construction in northwest, north and northeast China, where ecological conditions are quite fragile. The project includes reforestation and afforestation, water and soil conservation, windbreak and sand fixation. This project, including three main phases, covers an area of 4.07×10^6 km^2, including 13 provinces and municipalities (Fang, Chen, Peng, Zhao, & Ci, 2001; Yang, 2004; Wang, Innes, Lei, Dai, & Wu, 2007). According to the master plan, the total plantation will be 3.50×10^5 km^2, with 2.63×10^5 km^2 of artificial afforestation (75.1% of the land area), 7.60×10^5 km^2 of reforestation enclosure (21.7% of the land area), and 1.11×10^4 km^2 of airplane seeding (3.2% of the land area), to improve the forest coverage from 5.05 to 14.95% of the region. After more than 30 years of implementation, the total area of afforestation and reforestation is 2.45×10^5 km^2, an increase of 10.51% in forest coverage by 2010.

16.4.2 Natural Forest Protection

In the context of payments for ecosystem services, natural forest protection is classified into three different categories: major public benefits eco-forest, general public benefits eco-forest and commercial forestry bases (Zhang et al., 2000). There are two main zones for major public benefits eco-forest and key regions for natural forest protection: ① the upstream segments of the Yangtze River and the upstream and middle segments of the Yellow River; ② areas in northeast China and Inner Mongolia (Wang et al., 2007), where the natural forest covers 7.00×10^5 km^2, accounting for 69% of the national natural forest land. In these regions, the area of enclosure for reforestation is 3.67×10^4 km^2 (Yin & Yin, 2010), and the area of afforestation is 8.66×10^4 km^2, which will improve the forest coverage from 17.5 to 21.24%. The total investment in the project is ¥1.06×10^{11}, including 18.8% for infrastructure construction and 81.2% for afforestation, plantation and etc. In addition, according to the State Council, the financial debts of forestry businesses and enterprises seriously affected by logging reduction shall be deducted or exempted.

16.4.3 Beijing-Tianjin Sandstorm Sources Governance

The Beijing-Tianjin sandstorm sources governance project has been operating since 2000 and covers an area of 4.58×10^5 km^2, including Beijing, Tianjin, Hebei, Shanxi and Inner Mongolia (Yin & Yin, 2010), with an investment of ¥5.77×10^{10}. The purpose of this project is primarily to apply ecological engineering tasks, such as returning farmland to forest, reforestation and afforestation, grassland restoration, and small watershed governance, in the following areas: ① along the southern border of Otindag (Hunshandake) Sandy Land, Xilinguole, Inner Mongolia; ② along the north piedmont of Yinshan Mountain, Wulanchabu, Inner Mongolia; ③ along the border between Hebei and Inner Mongolia; ④ along the eastern border of Mu-Us (Maowusu) Sandy Land, Shanxi, forming a crescent-shaped ecological barrier to protect Beijing and Tianjin from damage during sandstorms. The area of reforested farmland is 2.63×10^4 km^2, the area of reforestation and afforestation is 4.94×10^4 km^2, the area of treated grassland is 1.06×10^5 km^2 and the area of integrated, treated small watersheds is 2.34×10^4 km^2. There are approximately 180,000 ecological immigrants who are compensated ¥5,000 each.

16.4.4 Returning Farmland to Forest/Grassland

The practice of returning farmland to forest was first initiated by former Premier Zhu in 1999, and has continued since then. In 2001, the practice of returning farmland to forest was officially included in the 10th FYP. According to the 10-year plan for returning farmland to forest, which covers 22 provinces and municipalities, the target area for returning farmland to forest is 5.30×10^4 km^2, the target area for reforestation and afforestation is 80.0×10^3 km^2, the target area for water and soil conservation is 3.60×10^5 km^2, and the target area for windbreak and sand fixation is 7.00×10^5 km^2 by 2010. The *Ordinance on Returning Farmland to Forest* was promulgated on December 14, 2002 by the State Council and explicitly stipulated the content, extent, measures, and responsibilities of this engineering project. Within this project, two types of stakeholders, farmers and local governments, will be compensated for their economic losses from returning farmland to forest/grassland. For farmers, food, seedling fees, and management and maintenance subsidies will be provided by the state. For local governments, the decrease in financial revenue should be compensated by financial transfer payments from the state.

16.4.5 Returning Rangeland to Forest/Grassland

This project was promulgated in December 2002 by the State Council and has been expanded since 2003 to cover 11 provinces and municipalities in the western region of China. During the first 5 years, 6.67×10^5 km^2 of rangeland, including desert grassland in west Inner Mongolia, Gansu and Ningxia, degraded grassland in east Inner Mongolia and north Xinjiang, and the river source district's grassland on the eastern Tibetan Plateau, accounting for 40% of the total of seriously degraded grasslands in western China, were under centralized treatment, such as complete enclosure, seasonal enclosure and grazing, or rotational grazing. During the implementation of this project, herders were compensated with food, grain, and fodder subsidies. Detailed compensation standards are listed in Table 16.4. By 2007, the total area of grassland enclosure was 3.46×10^5 km^2, including 1.64×10^5 km^2 of complete enclosure, 1.73×10^5 km^2 of seasonal enclosure and grazing, and 8.5×10^3 km^2 of rotational grazing. In addition, the area of newly seeded grassland was 6.5×10^3 km^2.

Since the late 1990s, the Chinese government has invested more than ¥7.00×10^{11} in large-scale ecological engineering projects, where more than ¥3.00×10^{11} was used for compensation payments. Summaries of the large-scale ecological engineering projects in China are shown in Table 16.4.

16.5 ECM for Sector

16.5.1 ECM for Forest

In China, forest is divided into two categories, public benefits eco-forest and commercial forest. The milestone of the development of ECM for forests was the establishment of the Forest Ecological Benefit Subsidy, along with the promulgation of

Table 16.4 Summary of large-scale ecological engineering projects in China

Ecological engineering	Started year	Area covered/10^6 km^2	Accomplishments by 2010/10^3 km^2	Major means of compensations
Three-north project	1978	4.069 (13 provinces and municipalities)	Reforestation and afforestation: 244.7 Water and soil conservation: 386.0 Windbreak and sand fixation: 278.0	No specific compensation standards were promulgated before 2000 The investments were from the central government for afforestation, airplane seeding, enclosure, grass seeds bases, pest control, cultivation management and promotion of plantation technologies
Natural forest protection	1998	(13 provinces and municipalities for Task 1) (5 provinces and municipalities for Task 2)	Reforestation: 36.7 Afforestation: 86.6	About 18.8% is for infrastructure construction and 81.2% comes from financial special funds Deduction or exemption of financial debts

(continued)

Table 16.4 (continued)

Ecological engineering	Started year	Area covered/10^6 km^2	Accomplishments by 2010/10^3 km^2	Major means of compensations
Beijing-Tianjin sandstorm sources governance	1999	0.458 (5 provinces and municipalities)	Returned farmland: 26.3 Reforestation and afforestation: 4.94 Treated Grassland: 106.0 Integrated treated small watershed: 23.4 Ecological immigrants: 180,000	Returning farmland (for 8 years): Grain compensation: ¥0.35/kg for 150,000 kg/(km^2·yr) One-time seeding fee for plantation: ¥75,000/km^2 Living subsidy: ¥30,000/(km^2·yr) Reforestation and afforestation Afforestation: ¥450,000/km^2 Airplane seeding: ¥180,000/km^2 Enclosure: ¥105,000/km^2 Grassland governance: Artificial grassing: ¥180,000/km^2 Airplane seeding: ¥150,000/km^2 Enclosure: ¥105,000/km^2 Pasture construction: ¥750,000/km^2 Grass seeds bases construction: ¥1,800,000/km^2 Fodder subsidy (for 5 years): ¥22.5/(kg·day·km^2) Small watershed governance: ¥200,000/km^2 Immigrants compensation: ¥5,000/person

(continued)

16.5 ECM for Sector

Table 16.4 (continued)

Ecological engineering	Started year	Area covered/10^6 km^2	Accomplishments by 2010/10^3 km^2	Major means of compensations
Returning farmland to forest/grassland	1999	(22 provinces and municipalities)	Returned farmland: 53.0 Afforestation: 80.0 Water and soil conservation: 360.0 Windbreak and sand fixation: 700.0	Eco-forest (for 8 years), economic forest (5 years): Grain compensation is ¥0.35/kg for 225,000 kg/(km^2·yr) (the Yangtze River) and 150,000 kg/(km^2·yr) (the Yellow River) One-time seeding fee for plantation is ¥75,000/km^2 Living subsidy is ¥30,000/(km^2·yr)
Returning rangeland to forest/grassland	2002	(11 provinces and municipalities)	Centralized treated rangeland: 670.0	Fodder subsidy (for 5 years): ¥0.225/kg in Tibetan Plateau: Complete enclosure: 4,125 kg/(km^2·yr) Seasonal enclosure: 1,035 kg/(km^2·yr) In Inner Mongolia, Xinjiang and Gansu: Complete enclosure: 8,250 kg/(km^2·yr) Seasonal enclosure: 2,063 kg/(km^2·yr) Fencing construction fee: In Tibetan Plateau: ¥37,500/km^2 Other places: ¥30,000/km^2 One-time seeding fee for grass plantation: ¥15,000/km^2

the Interim Measures of Forest Ecological Benefit Subsidy Management by the SFA and the MOF, jointly, in November 2001. The purpose was to set up a special fund for the protection, maintenance and management of the public benefits of eco-forests, including key shelter forests and special use forests. In October 2004, these policies were replaced by the Central Forest Ecological Benefit Compensation Fund and *the Measures of the Central Forest Ecological Benefits Compensation Fund Management*. In March 2007, the Central Finance Forest Ecological Benefit Compensation Fund (the Central Finance Compensation Fund) and the *Measures of the Central Finance Forest Ecological Benefits Compensation Fund Management* were promulgated to replace the policies enacted in 2004. According to the Central Finance Compensation Fund, the average compensation standard for a public benefit eco-forest is ¥7,500/(km^2 · yr) for a state-owned forest and ¥15,000/(km^2 · yr) for a collectively-owned or private forest. By the end of 2011, approximately 8.39×10^5 km^2 of public benefit eco-forest was compensated by the Central Finance Compensation Fund.

Based on different financial situations, several provincial governments have also set up local compensation funds for regional public benefits eco-forests. For example, in Beijing, the Fund for the Ecological Benefits Promotion and Development of Public Benefits Eco-Forest in Mountain Area was launched in 2010. The compensation standard is ¥6.0×10^4/(km^2 · yr), including ¥3.6×10^4 for compensation and ¥2.4×10^4 for operation and management, which is raised every five years. In Zhejiang Province, the Compensation System of Public Benefits Eco-Forest was established in 2004 with the compensation standard of ¥1.2×10^4/(km^2 · yr), which is raised every two years. By 2009, the compensation standard is ¥2.55×10^4/(km^2 · yr). In Fujian Province, the compensation mechanism for forest ecological benefits paid by downstream areas to upstream areas was initiated in 2007. From 2007 to 2009, the total paid compensation was ¥2.58×10^8, and since 2010, the compensation has been ¥2.58×10^8, each year.

As both the central and local ECMs for forest continued to improve, the extent of ecological benefit compensation was expanded to include all forests that are restored and established under the Returning Farmland to Forest and Natural Forest Protection programs (Li, Li, Li, & Liu, 2007), as shown in Table 16.5.

16.5.2 ECM for Coal Mining

Most of the coal mining (approximately 96%) in China is underground mining, where subsidence is the major environmental damage. Ecological compensation for coal mining is composed of three parts that are paid by coal mine contractors or developers. The first part is the compensation to direct victims of environmental damage, including agricultural crop compensation, attachment compensation and resident relocation, which are issued according to related regulations and standards. The second part is the cost of ecological restoration and environmental remediation, as well as administrative governance, which is the essential element of the ECM for coal mining to raise capital. For example, in Shanxi Province, the fee for ecological

16.5 ECM for Sector

Table 16.5 Ecological compensations for forest

Category	Function	Type	Principles of ECM
Public benefits eco-forest	Key shelter forest	Water source reserve forest	To raise the prices of water resources
		Water and soil conservation forest	Subsidy for plantation, tending, maintenance and management Compensations on ecological and environmental damage, especial for water and soil conservation
		Windbreak and sand fixation forest	Cash subsidy
		Shore protection forest	Land compensation and relocation subsidy
	Special uses forest	Defense forest	Subsidy for plantation, tending, maintenance and management
		Natural conservation forest	Compensation paid by beneficiaries of forest improvement
Returning farmland to forest	Ecological forest		Provide subsidy for 8 years Ecological benefit compensation fund Limited logging
	Economic forest		Provide subsidy for 5 years Flexible taxation policy for forest products
Natural forest protection	Natural forest		Subsidy for tending, maintenance and management Compensations to loss due to tree felling ban
	Artificial forest		Subsidy for plantation, tending, maintenance and management

compensation is ¥0.15 per ton of coal sold, which results in capital funds of ¥5.0 × 10^7 for reforestation and afforestation annually. The last ecological compensation fee is the tax for the sustainable development fund, which is ¥15/ton for thermal coal, ¥10/ton for anthracite coal and ¥15–20/ton for coking coal.

Based on the suggestions from Ma and Gao, the costs for ecological restoration, environmental remediation and administrative governance should be estimated according to local conditions, and the compensation fee should be collected by the fixed amount or fixed ratio methods, based on the estimated costs, as summarized in Table 16.6. However, environmental costs are severely underestimated because only subsidence is considered. Thus, neither the fixed amount nor the fixed ratio methods will be able to raise sufficient funds as needed. Due to the rapid depletion of fossil fuels, the average coal sale price has been sharply rising since 2008. Therefore, the fixed amount method might stimulate intentional or unintentional ignorance of environmental protection, driven by enormous profits. On the other hand, the fixed ratio might be a better approach to raise considerable funds for ecological compensation and restoration with higher sale prices. However, the ratio is too low to reflect the real market economy. It is necessary to elevate the ratio to a minimum of 3.5%. Additionally, a flexible ratio method should be more appropriate for a market economy.

Table 16.6 Ecological compensation for coal mining in various areas

Terrain	Compensations	S.E. coastal area	Bohai Rim	N.E. area	Mid area	S.W. area	Loess plateau	N.W. area
Plain area	Costs for ecological engineering/(CNY/ton)	1.76	1.46	2.07	2.08	2.12	1.78	1.28
	Costs for land reclamation/(CNY/ton)	5.28	4.37	6.21	6.24	6.36	5.33	5.35
	Taxation (percentage of sale price per ton[a])/%	2.20	1.82	1.59	2.60	2.65	2.22	2.23
Hills and mountains area	Costs for ecological restoration/(CNY/ton)	2.62	1.83	2.49	2.70	2.79	2.18	2.06
	Costs for land reclamation/(CNY/ton)	7.86	5.50	7.48	8.09	8.38	6.55	6.16
	Taxation (percentage of sale price per ton[a])/%	3.28	2.29	3.12	3.37	3.49	2.73	2.57

[a]The average coal sale price was ¥240/ton in 2008.

16.5.3 ECM for Grassland

As stipulated in the *Some Opinions Concerning Strengthening Grassland Protection and Construction*, promulgated by the State Council in September 2002, the government should provide food, cash and grass seed subsidy to farmers and herders for returning farmland to grassland. In addition, compensation and grassland vegetation restoration fees should be paid by those who utilize grassland, as regulated in Article 39 of the amended *Law of the People's Republic of China* on *Grassland* (the Grassland Law) in December 2002. Some compensation standards were also identified in the *Notice on Further Improving the Policy and Measures for Returning Farmland to Grassland*, promulgated by the NRDC, the MOA, the MOF and other government agencies jointly in April 2005, as shown in Table 16.4. Moreover, as projected in *the National Master Plan for Grassland Conservation, Construction and Utilizations*, (the Grassland Master Plan), promulgated by the MOA in April 2007, the total enclosed grassland will be 1.5×10^6 km^2, the total improved grassland will be 6.0×10^5 km^2 and the total artificial grassland will be 3.0×10^5 km^2 by 2020.

Based on the polluter pays principle, beneficiary pays and protector benefits, the ECM for grassland is implemented in four parts: project compensation, utilization compensation, encouragement compensation and other compensation.

(1) Project compensation

For most degraded grasslands, there are urgent needs for infrastructure construction, such as full-enclosure fence, artificial grassland planting, livestock pens, seed stock breeding, irrigation, transportation, and hazard prevention and mitigation. To restore degraded grassland, some projects are implemented as compensation to enhance the investment on grassland engineering projects to protect, construct and utilize grasslands. According to the Grassland Master Plan, there are 9 major engineering projects for grassland protection and construction, including returning rangeland to grassland, desertified grassland governance, grassland governance in the karst areas of southwest China, the fine grass and lawn seed incubation industry, hazard prevention and mitigation for grassland, engineering projects for grassland nature reserves, auxiliary projects on human-grass-livestock settlements for nomadic people, grassland utilization in agricultural areas and hydraulic engineering projects for pastoral areas.

(2) Utilization compensation

As legally and explicitly specified in Article 39 of the Grassland Law, utilization compensation is the payment for ecological restoration and environmental remediation, paid by developers, polluters and beneficiaries during the utilizations of natural grassland resources. Unfortunately, due to rapid changes in the monetized value resulting from rapid economic development and urbanization, there are no rational and sufficient implementation tools. For example, there are no appropriate compensation standards for ecological restoration, environmental remediation and suitable legal management schemes for fee collection to properly implement this policy.

(3) Encouragement compensation

As widely recognized, the most important factor that causes grassland ecosystem damage is overstocking. Therefore, to implement ecological restoration and environmental remediation, it is necessary to establish ECM programs that encourage a grassland-livestock balance. For example, the first program provides subsidies for the grassland-livestock balance. Based on the current grassland condition in China, the average grassland carrying capacity is 0.75 dry sheep equivalent (DSE is equivalent to 7.60×10^6 J/day) per hectare, the average economic profit is ¥200/DSE, and the average overstocking rate is 36%. Therefore, the subsidy will be ¥5,400/km^2 to reduce 36% of the grazing volume. The second program includes grasslands in the agricultural subsidy, providing subsidies for pasture seeds and artificial grass plantation.

(4) Other compensation

Several local governments have promulgated subsidy policies for local grassland ecological compensation, for example, subsidies that encourage herders to bring their livestock to market earlier to reduce the grazing pressure on grasslands during the winter season, subsidies for the construction of livestock pens, subsidies for machinery, subsidies for full enclosures and reduced grazing, and subsidies for immigrant relocation.

16.6 ECM at Local Level

16.6.1 ECM for the Three Rivers Source District, Qinghai Province

The Three Rivers (the Yangtze River, the Yellow River and the Lancang River) Source District (TRSD), the first National Comprehensive Pilot Site of Ecological Protection, is classified as a no exploitation zone according to the "major functional zoning strategy" promulgated through the 11th FYP. To compensate for the economic loss from not exploiting the resources, the ECM for the Three Rivers Source District has been jointly instituted and implemented by the central and local governments. As proposed in *the Master Plan for the Ecological Protection and Construction of National Nature Reserves at Three Rivers Sources District, Qinghai Province* (the Master Plan of TRSD), approved by the State Council in January 2005, the environmental protection and construction projects include returning rangeland to grassland, returning farmland to forest, ecological relocation, potable water supply, and water and soil conservation, with a total investment of ¥7.5×10^9. For local governments, in 2003, the Qinghai Province Government launched its "returning rangeland to grassland" program for natural grassland restoration. In this program, approximately 1.02 ×

10^4 km^2 of natural grassland were fully enclosed for restoration, with a total investment of ¥3.15 × 10^8. The fee for fence construction is ¥30,000/km^2, and the fodder subsidy (as stale rice) is 4,125 kg/(km^2 · yr) for 5 years.

16.6.2 ECM for Poyang Lake, Jiangxi Province

Poyang Lake, located in the middle Yangtze River, is the largest freshwater lake in China, which helps to regulate water in the Yangtze River by providing water to the river during low water periods and receiving water from the river during high water periods. As one of the seven most important wetlands in the world, Poyang Lake plays an important role in biodiversity protection, freshwater resource maintenance, flood control, climate regulation, pollution reduction, and products and life support for human beings. To preserve this water resource and the ecological environment of Poyang Lake and to protect the ecological safety of the Poyang Lake watershed and the middle and downstream regions the Yangtze River, the Poyang Lake Ecological Economic Zone was proposed in 2008 to realize regional sustainable development. It became a national strategy in 2009.

The ecological compensation for the Poyang Lake area, including national compensation, local government compensation, social compensation and internal compensation, can be characterized as (i) itemized compensation mechanisms that are implemented by the departments of agriculture, forestry, water resources and state land, according to individual policies that are promulgated based on *the Ordinances on Returning Farmland to Forest* (2002), the Water Law (2002), *the Law of the People's Republic of China on Mineral Resources* (1996), *the Law of the People's Republic of China on Fisheries* (2004), *the Ordinances on Wetland Protection Poyang Lake* (2003), and other national and local regulations; and (ii) vertical compensation mechanisms that are the financial transfer payments from the central government. However, most of the capital is used for ecological cultivation, maintenance and management, with less for farmers and herders. For example, the relocation compensation was only ¥33,700/house for people returning farmland to lake, and the subsidy for the fishing ban period (three months) was only ¥400 per fishing boat per year (Tang et al., 2009).

16.6.3 ECM for Nature Reserves, Hainan Province

Nature reserves play a significant role in water resource preservation, soil and water conservation, environmental quality improvement and the maintenance of ecological balance. To preserve the ecological functions and ecosystem services of nature reserves, protect biodiversity and achieve local sustainable development, ECMs for nature reserves were initiated in the Hainan Province in 2006. Under this system, ecological functions and ecosystem services are predominantly categorized as environmental quality improvement, forestry resources, water resources, aquaculture, mining, land resources, tourism resources and wildlife, and various ECM programs are applied, including wastewater treatment, forest resource protection, water conservation facility construction, fishery resource protection, ecological construction, wildlife protection and subsidies for returning farmland to forest. The major capital source for ECM implementation is payment for ecosystem services (PES) from private sectors for the utilization of ecological functions and ecosystem services, including emission fees, forestry resource fees, water resource fees, use tax for water bodies, mineral resource fees and land reclamation fees, higher prices for entrance tickets and penalties for violation. Among these ECM programs, only the compensation for returning farmland to forest is acquired through financial transfer payment by the government. The compensation for agricultural crops, i.e., paddy, is ¥825,000/(km$^2 \cdot$ yr), and the opportunity cost is ¥1,000/(house $\cdot$ yr).

16.6.4 The ECM for Natural Resource Utilization, Sichuan Province

In Sichuan, despite its abundant natural resources, economic development has not benefited from the exploitation and utilization of natural resources, due to insufficient ecological compensation, low resource taxes, minimum labor costs and underpriced resource products. For example, the subsidy for forest tending, maintenance and management is ¥1,950/(km$^2 \cdot$ yr) and the public benefits eco-forest subsidy is ¥7,500/(km$^2 \cdot$ yr). The resource tax is ¥8–30/ton for crude oil, ¥2–15/ton for natural gas, ¥0.3–5/ton for coal, ¥0.5–20/ton (or m^3) for nonmetallic minerals, ¥2–30/ton for ferrous metal minerals and ¥0.4–30/ton for nonferrous metal mineral. For labor cost, local people receive approximately ¥10/ton for excavated phosphorus ore, which is sold for ¥280/ton, on average, by the mine contractors. For resource products, the electricity price from Xiluodu Hydropower Station is only ¥0.3635/(kW $\cdot$ h), whereas the electricity price in Shanghai is ¥0.6/(kW $\cdot$ h).

16.6.5 The ECM for Watershed Conservation, Yunnan Province

The Songhua Dam Water Source Reserve is located in Qunming, Yunnan Province. Due to low water prices, the residents do not receive sufficient compensation because the majority of the compensation is allocated to infrastructure construction and minimum maintenance and management. Compensation is only provided for ecological

16.6 ECM at Local Level

Table 16.7 Compensation programs for watershed conservation in Yunnan Province

Program	Means of ECM
Production subsidy	Paddy and corn cultivation: subsidy is ¥30,000/(km² · yr) Returning farmland to forest: Water conservation forest: cash subsidy is ¥450,000/(km² · yr) (for 12 years) Economic forest: cash subsidy is ¥450,000/(km² · yr) (for 8 years) Tending, maintenance and management: subsidy is ¥30,000/(km² · yr) (for 5 years) Balanced fertilization: subsidy is ¥75,000/(km² · yr)
Living subsidy	Residents in the reserve: Fuel subsidy is ¥8/(person · month) Medical subsidy is ¥8/(person · yr) Residents working outside the Reserve: subsidy is ¥300/(person · yr) Students attending high schools or equivalent outside the Reserve: subsidy is ¥300/(person · yr)
Management subsidy	Long-term forest maintenance workers: subsidy is ¥200/(person · month) Cleaning workers: subsidy is ¥300/(person · month)

construction and water source protection, and other forms of compensation are lacking. Currently, there are generally three types of compensation, including production subsidies, living subsidies and management subsidies, as summarized in Table 16.7.

(1) Production subsidy

For paddy and corn cultivation, the subsidy is ¥30,000/(km² · yr). For returning farmland to forest, water conservation forest (for 12 years) and economic forest (for 8 years), the cash subsidy is ¥450,000/(km² · yr) and the subsidy for tending, maintenance and management is ¥30,000/(km² · yr) for 5 years. The subsidy for balanced fertilization is ¥75,000/(km² · yr).

(2) Living subsidy

For residents in the reserve, the fuel subsidy is ¥8/(person · month), and the medical subsidy is ¥8/(person · yr). For residents working outside the reserve, the subsidy is ¥300/(person · yr), and for students attending high schools or equivalent outside the reserve, the subsidy is ¥300/(person · yr).

(3) Management subsidy

For long-term forest maintenance workers, the subsidy is ¥200/(person · month). For cleaning workers, the subsidy is ¥300/(person · month).

16.7 Conclusions

In China, strong demands on GDP growth has led to rapid resource depletion and striking environmental degradation. With astonishing economic growth, the costs are immense. With increasing consciousness of the importance of ecological restoration and environmental remediation, ECMs have been implemented since the 1990s to set limits on development, mitigate conflicts between different stakeholders, and provide financial support for eco-friendly measures and actions. But continued research and study should emphasize methods that make ECM programs more feasible, sustainable, effective, efficient and executable based on the comprehensive national power, overall economic strength, disparities between regional conditions, diverse folk customs, and various resource values (i.e., production cost, renewing cost, environmental cost, alternative cost, and service cost). As proposed, the following recommendations are all possible directions for future research: ① institute related laws and regulations to enhance the legality of ECM and increase the liability of contractors for resource exploitation; ② raise the taxation rates to increase the available monetary funds; ③ integrate environmental assessment with ECM; ④ establish compensation evaluation standards or indices for different areas and resources; ⑤ adopt carbon sink or carbon emission strength as one of the criteria for ecological service appraisal; ⑥ utilize public participation as the major means to supervise ECM programs; ⑦ educate local residents about more suitable and applicable occupations to facilitate the sustainability of ECMs.

References

Baylis, K., Peplow, S., Rausser, G., & Simon, L. (2008). Agri-environmental policies in the EU and United States: A comparison. *Ecological Economics, 65*(4), 753–764.
Chang, I.-S., & Wu, J. (2011). Review on natural resources utilization in China. *Management Science and Engineering, 5*(2), 12.
Fang, J., Chen, A., Peng, C., Zhao, S., & Ci, L. (2001). Changes in forest biomass carbon storage in China between 1949 and 1998. *Science, 292*(5525), 2320–2322.
Li, W., Li, S., Li, F., & Liu, M. (2007). Discussions on several issues of forest eco-compensation mechanism. *China Population, Resources and Environment, 17*(2), 6.
Pagiola, S., Arcenas, A., & Platais, G. (2005). Can payments for environmental services help reduce poverty? An exploration of the issues and the evidence to date from Latin America. *World Development, 33*(2), 17.
Peng, X. (2011). China's demographic history and future challenges. *Science, 333*(6042), 581–587.
Rondeau, D., & Bulte, E. H. (2007). Wildlife damage and agriculture: A dynamic analysis of compensation schemes. *American Journal of Agricultural Economics, 89*(2), 490–507.
Tang, M., Zhong, D., Ding, S., Peng, Y., Yu, T., & Xiao, J. (2009). Poyang lake area construction of a new mechanism for ecological compensation. *Guangdong Chemical Industry, 11*(39), 2.
Wang, G., Innes, J. L., Lei, J., Dai, S., & Wu, S. W. (2007). China's forestry reforms. *Science, 318*, 2.
Yang, H. (2004). Land conservation campaign in China: Integrated management, local participation and food supply option. *Geoforum, 35*(4), 507–518.

References

Yin, R., & Yin, G. (2010). China's primary programs of terrestrial ecosystem restoration: Initiation, implementation, and challenges. *Environmental Management, 45,* 13.

Zhang, P., Shao, G., Zhao, G., LeMaster, D. C., Parker, G. R., John, J., et al. (2000). China's forest policy for the 21st century. *Science, 288,* 2.

Chapter 17
Air Quality Governance in China

17.1 General Features of Air Pollution

China, as the biggest developing country, has been experiencing rapid industrialization and drastic urbanization, since the Reform and Opening-Up initiated in the late 1970s, gradually resulting in air pollution nationwide, just like all other developed countries. Along with huge economic growth, speedy urban expansion and modernization, and greatly improved living conditions, the needs for widespread infrastructure construction of transportation networks have sharply increased, resulting not only in the utilization of various kinds of vehicles, both in the private and public sectors, but also in the worsening of the problem of air pollution. In addition, the utilization of fossil fuels, especially coal as the major energy source, has exacerbated the issue of air pollution, aside from industrial emissions. However, due to huge differences in geographical locations, natural conditions, economic structure and developing modes between various regions, the type, extent, and the level of air pollution are quite diverse and can be characterized as regionalized, mingled, and complicated (Li, 2014; Du, Yao, Li, & Lan, 2015).

Actually, ambient air quality protection has always been one of the primary missions for the Central Government. Many policies, laws, regulations, ordinances, and measures related to air pollution prevention and control were promulgated by the MEP jointly with various government agencies, and have been revised, adjusted, and perfected several times along with the advancement of economic, social, and technological development. Indeed, these policies, laws, regulations, ordinances, and measures did play essential roles in air pollution prevention and control to achieve some remarkable accomplishments. Nevertheless, the status of ambient air quality still cannot be taken lightly, as there are some deficiencies and drawbacks within these policies, laws, regulations, ordinances, and measures.

Traditionally, air quality governance mainly relies on a local competent authority for environmental protection within the administrative jurisdiction, due to the great differences in geographical conditions, economic development and air pollution patterns between various regions. Due to the increasing complexity and diversity of contemporary air pollution, air quality governance has become a great challenge for the local competent authority for environmental protection to fight against, single-handedly. In addition, as the far-reaching effects of global warming and climate change gradually spread out, the issue of air quality is even more challenging and surpasses the capacity of traditional mode of air quality governance. Furthermore, based on the separation of duties, it is quite difficult for every government agency to fulfill its duties.

It was recognized that we should learn from past mistakes committed by western industrialized and developed countries to avoid the path of "pollution first and governance followed." Yet, this strategy was not well implemented. Ever since the Reform and Opening-Up, in the context of economic development first, environmental protection was seriously compromised by the need for economic growth to exacerbate the condition of air pollution, where the development of air pollution has gone through several stages, from single pollutant (soot) to multiple pollutants (soot, acid rain, and sulfur dioxide), and then to compound pollutants (soot, acid rain, haze, photochemical pollutants, toxic and hazardous pollutants), to significantly increase the complexity and difficulty of air quality governance for both central and local governments. In addition, the scale of air quality governance has evolved from a local scale to a local-regional scale, then to a regional scale, and finally to a cross-regional scale. Therefore, the gradually emerging consensus is that, in order to tackle these problems effectively and efficiently, multilateral collaboration between various government agencies and local governments should be deployed. Consequently, in order to seek mutual benefits of various provinces and cities in a more effective way within a shorter period of time, several cross-regional collaborations were initiated, such as the Jing-Jin-Ji Area (including Beijing, Tianjin, and Hebei), the Triangle Area of the Yangtze River Delta (including Shanghai, south Jiangsu, and north Zhejiang), and the Triangle Area of the Pearl River Delta (including nine major cities in Guangdong) to facilitate regional air quality governance through experience exchange and resources sharing.

17.2 Legal Institution and System

The legal institution and system of air pollution protection in China refer to laws, regulations, ordinances, directives, decrees, rules, provisions, planning, measures, etc. promulgated for air pollution prevention and control and performance evaluation, as well as the standards of energy conservation and emission reduction, which are the foundations of air quality governance to facilitate the implementation of policies, to evaluate the progress and to assess the outcomes. Generally, the laws of air pollution prevention and control can be categorized into the following parts (Wang & Wang, 2014; Du, Yao, Li, & Lan, 2015).

17.2 Legal Institution and System

(1) Foundation

All legal institutions and systems are based on the Constitution which is the fundamental law of the State. As specified in Article 26 of the Constitution, the State shall protect and improve the living and ecological environment, and prevent and control all kinds of pollution and other public health hazards, thus laying the foundation of legislation for air pollution prevention and control.

(2) Legal framework

The framework of a legal institution and system for air quality governance was constructed based on the EP Law and the Air Pollution Law. The obligations of government agencies on the supervision and management for air pollution prevention and control, management of emission permit for the private sector, countermeasures for various kinds of pollution, legal liability of enterprises leading to air pollution, etc. were all clearly stipulated in these laws to formulate the legal framework.

(3) Supplement

The supplement of the legal institution and system for air quality governance is mainly composed of the *General Provisions of the Civil Law of the People's Republic of China*, the *Law of the People's Republic of China on Liability for Tort*, the *Criminal Law of the People's Republic of China*, etc. in which the civil liability and criminal liability of perpetrators causing the air pollution are distinctly specified in a different form from that of the administrative liability set forth in the legal framework.

(4) Guidance

The guidance of the legal institution and system for air quality governance includes some departmental regulations, including the *Interim Measures of Developing Coal Briquette for Civilian Applications* (July 12, 1987), the *Measures of Management for Smoke and Dust Control in Cities* (July 21, 1987), the *Measures of Management for Supervision on Motor Vehicle Exhaust Pollution* (August 15, 1990), and several key normative documents, such as the *Norm of Ambient Air Quality Monitoring (on Trial)* (January 19, 2007), the *Amendment of Ambient Air Quality Standards* (February 29, 2012), the *Action Plan for Air Pollution Prevention and Control, the Compilation of Advanced Technologies of Air Pollution Prevention and Control* (March 3, 2014), etc., promulgated by the State Council and the MEP, to provide necessary guidelines to air quality monitoring, measures and techniques for pollution prevention and control, and other aspects of air pollution prevention and control.

(5) Detailed implementation rules

Detailed rules are largely comprised of regional and local normative documents issued by regional and local governments as air quality governance would be implemented locally with support from regional and local governments and collaboration from the private sector. All regional or local normative documents were published as per regional conditions and local needs; for example, the Ordinances of Air Pollution Prevention and Control, issued in Beijing, Jinan, Nanjing, Wulumuqi and Shaanxi province; the Regulations of Air Quality Prevention and Control Management, promulgated in Shijiazhuang, Zhuhai and Chengdu; and the Liability Agreement on the Objectives of Air Pollution Prevention and Control, signed by all 31 provinces, autonomous regions, and municipalities directly under the central government with the MEP, on behalf of the State Council. In summary, since entering the 21st century, more than 20 documents concerning air pollution prevention and control, including laws, regulations, policies, technical guidelines, directives, standards, etc., were promulgated, as listed in Box 17.1. Many detailed implementation plans, guiding schemes, pollution prevention and control programs were also published, for example, the Two-Control-Zone, as shown in Box 17.2.

Box 17.1 Some documents concerning air pollution prevention and control since the 21st century

- *the Amendment of the Air Pollution Law (April 2000).*
- *the Policy on Pollution Prevention and Control Technologies of Sulfur Dioxide Emission from Coal Burning (January 2002).*
- *the Ordinance of Collection, Utilizations and Management for Emissions charges (January 2003).*
- *the Amendment of the Air Pollutants Emission Standards for Thermal Power Plant (December 2003).*
- *the Technical Guidelines for Air Pollution Governance Engineering (December 2010).*
- *the Amendment of the Air Pollutants Emission Standards for Thermal Power Plant (July 2011) and the Policy on Integrated Pollution Prevention and Control Technologies for Fine Particulate Matter (September 2013).*
- *Acid Rain Control Zone and Sulfur Dioxide Pollution Control Zone during the 10th FYP (October 2002), the Schemes of Designating Key Cities for Air Pollution Prevention and Control (December 2002), the Measures of Management for Electricity Price for Coal Fired Power Generator with Desulfurization Facility and the Operation of Desulfurization Facility (on Trial) (May 29, 2007), the Guiding Opinions Concerning the Promotion of Unified Regional Air Pollution Prevention and Control Work for Improving Regional Air Quality (May 2010), the Air Pollution Prevention and Control for Key Regions during the 12th FYP (October 2012), the Action Plan for Air Pollution Prevention and Control (September 2013), and the Action Plan for Energy Conservation, Emission Reduction and Low Carbon Development between 2014 and 2015 (May 2014).*

> **Box 17.2 Several plans, schemes and programs of air pollution prevention and control since the 21st century**
> - *Acid Rain Control Zone and Sulfur Dioxide Pollution Control Zone during the 10th FYP (October 2002).*
> - *the Schemes of Designating Key Cities for Air Pollution Prevention and Control (December 2002).*
> - *the Measures of Management for Electricity Price for Coal Fired Power Generator with Desulfurization Facility and the Operation of Desulfurization Facility (on Trial) (May 29, 2007).*
> - *the Guiding Opinions Concerning the Promotion of Unified Regional Air Pollution Prevention and Control Work for Improving Regional Air Quality (May 2010).*
> - *the Air Pollution Prevention and Control for Key Regions during the 12th FYP (October 2012).*
> - *the Action Plan for Air Pollution Prevention and Control (September 2013).*
> - *the Action Plan for Energy Conservation, Emission Reduction and Low Carbon Development between 2014 and 2015 (May 2014).*

17.3 Process of Development

Along with rapid economic growth and drastic social reorganization, the rapid transformation and transition in all aspects resulting from the Reform and Opening-Up have generated huge challenges and demands for change for policy makers due to inappropriateness, insufficiencies, and disadvantages within the existing policies and institutions of air quality governance. Many policies, laws, regulations, provisions, ordinances, etc. were amended and revised several times according to the conditions of environmental quality and the status of economic and social development; for example, the Air Pollution Law was initially promulgated on September 5, 1987, with the several amendments on August 29, 1995, April 29, 2000, August 29, 2015, and October 26, 2018. Generally, the development of legal institution and system of air pollution prevention and control can be categorized into three distinct phases: Phase I (from 1979 to 1997), Phase II (from 1998 to 2012) and phase III (from 2013 to date), as illustrated below (Mu & Zhang, 2013; Liu, 2016).

17.3.1 Phase I: Legal Institution Management for Air Pollution Control on Soot (from 1979 to 1997)

Since the commencement of the Reform and Opening-Up, environmental problems were becoming more prominent the energy demands began increasing drastically due to rapid industrial and economic development. The promulgation of the *Law of the People's Republic of China on Environmental Protection (on Trial)*, the EP Law (on Trial), on September 13, 1979 laid the legal groundwork for air pollution governance to mark the chapter of legal institution management for air pollution prevention and control. The key features of air pollution control during this period could be summarized as follows: focus on air pollution control over industrial point sources; soot is the major pollutant for air pollution control; the extent of pollution control is limited to locale; and the mode of governance is typically regionalism management, typically.

(1) The initiation of the legal institution of pollution prevention and control

The legal institution of pollution prevention and control was initiated by the promulgation of the EP Law (on Trial), in which it was clearly stipulated that all smoke emission units, industrial furnaces, ships, and motor vehicles shall install effective smoke abatement and dust collection devices, and the emissions of hazardous gases shall not exceed national standards. According to the EP Law (on Trial), the *Ambient Air Quality Standards* were published on April 06, 1982 to further expand the scope of air pollution control over six different pollutants, including total suspended particles (TSP), particulate matter (PM), sulfur dioxide (SO_2), carbon monoxide (CO), nitrogen oxides (NO_x), and ozone (O_3), respectively. As specified in the Air Pollution Law (September 05, 1987), all relevant competent authorities of the State Council shall stipulate strictly quality standards for boilers to limit the emissions of hazardous exhaust and dust into ambient air from boilers, according to the related national standards concerning dust emission for boilers. Moreover, more detailed and concrete operation guidance was provided through the promulgation of the *Implementation Provisions for the Law of the People's Republic of China on Air Pollution Prevention and Control* (the Implementation Provisions for the Air Pollution Law) on May 08, 1991. In addition, the first batch of standards for motor vehicle exhaust was publicized in 1983 as the first regulation to impose an emission limit on motor vehicle exhaust. In addition, as specified in the *Measures of Management for Supervision on Motor Vehicle Exhaust Pollution* promulgated August 15, 1990, the competent authority for environmental protection at all levels shall be responsible for the unified supervision and management for motor vehicle exhaust pollution between various departments.

(2) The commencement of environmental economy policy

The commencement of the emissions charges system was the first application of environmental economy policy to pollution control. It was originally proposed during the Fourth Meeting of the LGEP of the State Council on December 31, 1978, and

17.3 Process of Development

then firmly endorsed in the EP Law (on Trial). As the promulgation of the *Interim Measures of Emissions Charges Collection* promulgated on February 5, 1982 by the State Council, the emissions charges system was officially established. In order to control the increasingly deteriorating situation of sulfur dioxide pollution and acid rain damage, the *Notice on the Implementation of the Pilot Project to Collect Emissions Charges of Sulfur Dioxide from Industrial Coal Burning* was published on September 14, 1992 to further solidify the application of the emissions charges system on air pollution prevention and control. Moreover, the scale of pilot work for collecting emissions charges of sulfur dioxide was enlarged as the dissemination of the *Rescriptum on the Issues Concerning the Expansion of the Pilot Project to Collect Emissions Charges of Sulfur Dioxide* on April 2, 1996. Next, a pilot project of emission trading was deployed, which originated in the United States and introduced into China in the 1990s for acid rain control. In 1991, 16 cities were selected to employ the pilot work of emission permits for air pollutants. In 1994, six cities, Baotou, Kaiyuan, Liuzhou, Pingdingshan, Taiyuan and Guiyang, were designated to implement the pilot program of emission trading for air pollutants. In the same year, *the Outlines of National Environmental Protection Work from 1993 to 1998* was passed during the National Environmental Protection Work Meeting to clearly specify the continuation of the pilot program of emission trading for air pollutants.

(3) The implementation of regionalism management governance

The regionalism management governance mode was incarnated in administrative jurisdiction for stationary sources control and pollution prevention and control plans. As stipulated in Article 2 of the Air Pollution Law, the State Council and local governments at all level shall integrate the work of ambient air protection into the plans for national economic and social development, including the rationalization of industrial layout, enhancement scientific research on air pollution prevention and control, and adoption of necessary measures for air pollution prevention and control. Furthermore, as specified in Article 3, the State shall take essential measures to prepensely control or gradually cut down the total emission of major air pollutants in various areas and to be responsible for air quality within their jurisdictions; the local governments at all levels shall make plans and take measures to ensure the air quality within their jurisdictions are compliant with relevant standards. Thus, clearly, the mode of air quality governance is regionalism management governance, based upon administrative jurisdiction, where the Central Government is responsible for the overall planning for air quality of the entire country, and local governments are responsible for local air quality governance. However, there are some limitations and defects within this mode, which will be depicted below.

17.3.2 Phase II: Exploring Unified Regional Air Pollution Prevention and Control for Emerging Compound Pollution (from 1998 to 2012)

In 1998, the Two-Control-Zone (acid rain control zone and sulfur dioxide pollution control zone) was ratified by the State Council to officially initiate a new era for air pollution prevention and control. The rapid upsurge in sulfur dioxide emission from coal burning on account of rapid industrial and economic development led to disastrous climate conditions, such as acid rain, which resulted in severe damage. For instance, SO_2 emission in 1995 was 23.7 million tons, the highest in the world, surpassing even that of the European Union and the United States. The key air pollutants for air pollution control were sulfur dioxide, nitrogen oxides, and particulate matter. In addition, motor vehicle exhaust control was also included as one of the major tasks. Through more stringent air quality standards and tougher measures of urban air pollution prevention and control, China's achievements of urban air quality governance were widely recognized by the international community, especially during the 2008 Summer Olympics (Beijing), the 2010 World Expo (Shanghai), and the 2010 Asian Games (Guangzhou). The key features of air pollution control during this period of time could be summarized as follows: the Two-Control-Zone strategy was promulgated by responding to the fact that acid rain and sulfur dioxide pollution were in transition from local stationary sources pollution to regional pollution to demonstrate the characteristic of regional and compound pollution, so that the governance mode had to be changed from the local administrative jurisdiction management into the unified regional management.

(1) The Two-Control-Zone and regional supervision and inspection

The Two-Control-Zone was first proposed in the first Amendment of the Air Pollution Law (1995). As depicted in Article 27, for areas damaged or most likely to be damaged by acid rain, and areas seriously contaminated by sulfur dioxide pollution, the acid rain control zone and sulfur dioxide pollution control zone should be designated by the competent authority for environmental protection and other relevant departments of the State Council, according to the natural conditions, including climate, topography, geography, and soil, and ratified by the State Council. However, no detailed provisions concerning concrete legal liability and punishment were clearly stipulated. On January 12, 1998, the Two-Control-Zone was formally established through the promulgation of the *Zoning Schemes of the Acid Rain Control Zone and Sulfur Dioxide Pollution Control Zone*, in which the area of acid rain control zone was 800,000 km^2 and the area of sulfur dioxide pollution control zone was 290,000 km^2 to cover more than 175 cities. Then, as specified in the second Amendment of the Air Pollution Law (2000), a total emission cap system was implemented for key polluted areas with strict emission limits. Furthermore, it was proposed that a deadline be set for achieving the plan of air pollution control for major polluted cities. As required by the Two-Control-Zone during the 10th FYP, the total sulfur dioxide emission for the Two-Control-Zone in 2005 shall be reduced by 20% compared to that in 2000, in which a series of countermeasures for pollution prevention and control were proposed, such as reducing the sulfur content in coal, controlling the sulfur dioxide emission from coal-fired power plants, industrial furnaces, manufacturing processes, and so on.

In order to effectively supervise, inspect, and coordinate the air quality governance work of local governments and local competent authorities for environmental protection, six centers for regional supervision and inspection, namely, east, south, north, southwest, northwest, and northeast, were successively established from 2002 to 2008, to lay the foundations of unified regional air pollution prevention and control, and to explore the feasibility of vertical administrative management for local environmental protection work. Through trans-boundary unified management and cross-departmental collaboration on comprehensive control over various air pollutants, such as sulfur dioxide, nitrogen oxides, particulate matter, volatile organic compounds, etc., outstanding accomplishments were concretely achieved during the 2008 Summer Olympics, the 2010 World Expo, and the 2010 Asian Games to further demonstrate the importance and effectiveness of unified regional air pollution prevention and control. However, without a long-term regional cooperative mechanism, the continuous improvement and optimization of regional air quality cannot be assured, if only relying on some temporary schemes adopted for special events. Therefore, as pointed out in the *Guiding Opinions Concerning the Promotion of Joint Regional Air Pollution Prevention and Control Work for Improving Regional Air Quality*, promulgated jointly by the MEP, NDRC, Ministry of Science and Technology (the MST), etc., on May 11, 2010, it is necessary to implement the approach of unified regional air pollution prevention and control, as soon as possible, to resolve the issue of regional air pollution. It was the first time that the approach of unified regional air pollution prevention and control was proposed to be implemented nationwide.

(2) The introduction of market and competitive mechanisms

In their pursuit of better economic development and higher personal profits, a few of local governments and private enterprises are reluctant to actively undertake the tasks of air quality governance, due to the externality, resulting in the "Tragedy of the Commons." In a market economy, a competitive mechanism will be favorable for breakthroughs in dilemma and conflicts resolution. In conforming to the development of a market economy, a market competitive mechanism was introduced in air pollution prevention and control during this period. First, the extent of taxation on emissions charges of sulfur dioxide was widely expanded; consumption tax on leaded gasoline was adjusted; and market competitive mechanism was initially introduced into the market of EIA consulting services through pilot programs. Then, emission permit system was established; emission standards of motor vehicle exhaust were elevated; and the emissions charges rates were raised. As two important documents, the *Notice on the Demonstration Work Arrangement for Total Emission Cap for Sulfur Dioxide and Emission Trading Policy* and the Two-Control-Zone during the 10th FYP, were promulgated in May 2002, the investment on the pollution prevention and control for the Two-Control-Zone was increased, the execution of emissions charges of sulfur dioxide was enhanced, preferential policy for sulfur dioxide emission control was promoted, and seven provinces and cities, including Shandong, Henan, Shanxi, Jiangsu, Shanghai, Tianjin and Liuzhou, were selected for pilot projects of sulfur dioxide emission trading system. As the enactment of the *Ordinances of Management for the Collection and Utilizations of Emission Charges* on January 2, 2003, the emissions charges were paid according to the type and amount of discharged pollutants.

(3) The institution of the development strategy of energy conservation and pollution reduction

Overuse of fossil fuels is the primary cause of sulfur dioxide and nitrogen oxides pollution in ambient air. As a big country of coal reserve and consumption, coal resource plays the significant role in China's national economic development. Along with the continuous growth in the living standards, the number of private motor vehicles increases drastically to induce higher and higher gasoline consumption, which is unfavorable for air quality governance. As energy demands continually increasing, energy conservation and pollution reduction have become the significant measures and primary requirements for environmental protection, in conforming to the principles of sustainable development, which has become one of China's national policies since 1992. In addition, the principles of sustainable development have been gradually incorporated into China's energy policies. In addition, mitigating the environmental impacts resulted from energy consumption and promoting the development of renewable energy were stipulated in many environmental protection laws and regulations. As specified in the *Law of the People's Republic of China on Energy Conservation* (the Energy Conservation Law), promulgated on November 1, 1997, it is essential to promote the utilizations of renewable energy, to facilitate energy conservation in society at large, to enhance the efficiency of energy consumption, and to protect and improve the environment. In addition, some concrete measures of energy conservation for various sectors were proposed, such as industry, building construction, transportation, public institutions, and major energy consumption facilities, to further establish the development strategy of energy conservation.

17.3.3 Phase III: Integrating Unified Regional Air Pollution Prevention and Control for Compound Pollution (from 2013 to Date)

As rapid industrialization and comprehensive integration of regional economy, the characteristics of air pollution gradually became compound and regionalized, and the effects of the Two-Control-Zone were limited. Triggered by the recurrence of heavy haze episode during January 2013, which has caused huge direct economic loss (more than $3.6 billions) and severe public health damage (more than $3.5 billions), the most stringent measures of air pollution prevention and control were then promulgated. On September 10, 2013, the Action Plan for Air Pollution (or,

17.3 Process of Development

the Ten Tactics of Air Pollution) was officially disseminated by the State Council to clearly expound the detailed procedures of these ten tactics (35 articles), as shown in Box 17.3. The key features of air pollution control during this period could be summarized as: more rigorous laws and regulations were enacted in order to resolve the increasingly deteriorating air quality, especially the heavy haze episode, in addition to flexible market mechanisms; and unified regional air quality management was further confirmed through prompt legislative processes.

> **Box 17.3 Ten tactics of the Action Plan for Air Pollution Prevention and Control**
> - *Tactic-1: To strengthen integrated governance and to reduce various pollutants emissions.*
> - *Tactic-2: To adjust and optimize industrial structure and to promote the transformation and upgrade of industrial structure.*
> - *Tactic-3: To expedite the industrial technology reformation and to enhance the technological innovation.*
> - *Tactic-4: To advance the energy structure adjustment and to increase the supply of clean energy.*
> - *Tactic-5: To reinforce the qualification of energy saving and environmental protection, and to optimize the layout of industrial distribution.*
> - *Tactic-6: To utilize the effects of market mechanism and to perfect the environmental economics strategies.*
> - *Tactic-7: To improve the legal system and institution, and to intensify the supervision and management.*
> - *Tactic-8: To establish the mechanism of regional collaboration and to integrate regional environmental governance.*
> - *Tactic-9: To found the monitoring, pre-warning and emergency response system, and to counteract the heavy pollution weather.*
> - *Tactic-10: To clearly demarcate the responsibilities of the governments, the enterprises and the societies, and to mobilize the public participation in environmental protection.*

(a) To promulgate more stringent tactics for air pollution prevention and control

Comprehensive and concrete measures were specifically stipulated in 10 Articles (35 Terms) of the Action Plan for Air Pollution, known as the most powerful and stringent measures for air pollution prevention and control in Chinese history, to include: regulating multi-pollutants emission; promoting Industrial optimization and upgrade; expediting technological transformation of enterprises; advancing energy structure adjustment; optimizing the spatial layout of industries; perfecting the environmental economy policies; improving the laws and regulations system; instituting regional coordination mechanisms; establishing monitoring, and early warning and response system; and elucidating the liabilities of governments, enterprises and society at large. Various transformations and breakthroughs of China's air pollution prevention and control work were solidly embodied in the Action Plan for Air Pollution. First, the objective was transited from sole total emission cap to holistic air quality improvement. Second, the target was transformed from single pollutant control to integrated control over multiple compound pollutants. Third, the approach was changed from pure technological and technical means to manifold means, such as legal, economic, social supervision means, and so on, in addition to technological and technical means. Last, the mode was switched from local administrative jurisdiction management to the integration of local administrative jurisdiction management with unified regional management. In order to ensure the Action Plan for Air Pollution to be implemented smoothly and successfully, the Liability Agreement on the Objectives of Air Pollution Prevention and Control was signed by all 31 provinces, autonomous regions and municipalities directly under the central government with the MEP, on behalf of the State Council, in which concrete targets and tasks for each party were clearly specified, the executing schemes of the Action Plan for Air Pollution were proposed, and all targets and tasks were disintegrated into sub-targets and sub-tasks for primary levels for each year. On July 10, 2012, *the Special Plan for the Science and Technology of Blue Sky Project during the 12th FYP* was promulgated by the MST and the MEP, jointly, to provide scientific and technological supports to improving the environmental quality, to enhance energy conservation and emission reduction, and to guide the development of energy conservation and environmental protection industries. Moreover, on January 14, 2013, *the Notification Concerning further Perfecting the Air Quality Monitoring and Early Warning Work under the Weather Conditions for Potential Heavy Air Pollution* was promulgated by the MEP to target fine particulate matter as the key pollutant, to enhance the leadership, and to clarify the liability for air pollution prevention and control.

(b) To emphasize market mechanisms for air pollution prevention and control

Many supporting and supplementary polices were instituted to ensure the implementation of the Action Plan for Air Pollution, in which market incentives have been integrated as the mechanism to promote air pollution prevention and control. First, in order to promote the production and supply of oil fuels conforming to the Phase-IV National Standards for Motor Vehicle Exhaust, the principle of higher quality for higher price was applied to increase the quality of oil products. Second, in order to fully embody the integration of policy guidance and market incentives, various incentive policies were publicized by the Ministry of Finance (the MOF) to promote the development and deployment of "new energy" cars (excluding gasoline and diesel), while the special fund "Substituting Subsidies with Rewards" was established to encourage any entities with outstanding contributions to ambient air protection. Third, in order to actively promote the development of renewable energy and optimize the pricing structure of electricity, a strategy was employed by the NDRC to adjust the prices of electricity generated from renewable energy. As pointed out during the Third Plenary Session of the 18th CCCPC, the market mechanism shall play a decisive role in resource allocation. In addition, as indicated by both international and domestic practical experience, the market mechanism is an effective measure for the promotion of air pollution prevention and control. Therefore, in addition to policy guidance for air pollution prevention and control, the market mechanism of environmental protection shall be further improved to facilitate the effective integration of policy guidance and the market mechanism.

(c) To promote legislation for unified regional air pollution prevention and control

As explicitly stipulated in the *Air Pollution Prevention and Control for Key Regions during the 12th FYP*, promulgated by the MEP on October 29, 2012, regional compound air pollution has become a huge challenge to the current environmental governance mode, which mainly relies on administrative jurisdiction management to focus exclusively on air pollution prevention and control for a single city; in the context of increasing deterioration of air quality, it is necessary to establish a brand new regional air quality governance system. Furthermore, as required, a leading team of those involved in unified regional air pollution prevention and control work, summoned by the MEP with relevant departments and provincial governments with the region, shall be established to host regular meetings for sharing their experiences and setting work targets, key points and major tasks for the next period.

On August 29, 2015, the Amendment of the Air Pollution Law (2015) was promulgated to incorporate major revisions in eight chapters containing 129 articles. The Action Plan for Air Pollution was effectively transformed into a legal institution within the Air Pollution Law (2015). In addition to the general principles, legal liability, and supplementary provisions, the standards of air pollution, plans for attainment deadline, supervision and management, measures of air pollution prevention and control, unified regional air pollution prevention and control for major areas, and response to heavy pollution were also unequivocally regulated in the Air Pollution Law (2015). One of the major highlights of this Amendment is the integration of the unified regional air pollution prevention and control into a legal document in which it is mentioned in the general principles; moreover, a special chapter, the unified regional air pollution prevention and control for major areas, was instituted to clearly ascertain the goal of improving air quality, to propose a comprehensive and unified regional air pollution prevention and control for various industries, and to raise the requirement of collaborative control over multiple pollutants. Furthermore, the usage of the terms, the "Two-Control-Zone" and the "Key Cities for Air Pollution Prevention and Control", was both stopped, and the boundary of the control zone extended to the entire nation, to denote a great leap in the legislative process for unified regional air pollution prevention and control.

17.4 Evaluations and Prospects

The issue of air pollution in China has been gradually attracting public attention since the Reform and Opening-Up. Even though, it was strongly emphasized that international experience shall be used as a reference to avoid their mistakes, the growth of air pollution in China showed that it was basically repeating the same mistakes. Unfortunately, due to the rapid and drastic transformation of the characteristics of air pollution, the severity and extent of air pollution in China is even worse than that ever encountered by the industrial and developed. Therefore, through a trial-and-error process, the development of air pollution prevention and control in China has gone through many different stages, as described in detail above. Indeed, many remarkable achievements were gradually realized through continuous practice and enhancement, and there are some aspects need to be improved. (Li, 2014; Luo, 2014; Wang & Wang, 2014; Liu, 2016).

References

Du, P., Yao, J., Li, Y., & Lan, Y. (2015). Assessment and recommendations of air pollution control policy of China. *Sichuan Environment, 34*(1), 96–100.

Li, N. (2014). *The regional legal system for the prevention and control of atmospheric pollution linkage studies in China.* Master: Jiangxi University of Science and Technology.

Liu, X. (2016). *Evolution and optimization of air pollution control policy in China since the reform and opening up.* Master: Hunan Normal University.

Luo, E. (2014). Status and countermeasures of legislation for air pollution prevention and control in China. *Legal System and Society, 9,* 257–260.

Mu, Q., & Zhang, S. (2013). An evaluation of the economic loss due to the heavy haze during January 2013 in China. *China Environmental Science, 33*(11), 2087–2094.

Wang, T., & Wang, L. (2014). Analysis and optimization of China's air pollution laws. *Resources and Industries, 16*(2), 32–36.

Chapter 18
Water Quality Governance in China

18.1 Water Resources in China

Water is one of the essential resources that has nurtured human civilization throughout history and is one of the key factors that affects regional economic growth, social development, and environmental supporting capacity. In China, led by the Yangtze River, there are seven major watershed systems. Though China has abundant water resources that account for 6–7% of the total water resources in the world, water resources per capita is about 2,100 m^3, which is only equivalent to 25% of the world average; and water resources per area is about 2.99×10^5 m^3/km^2, which is 83% of the world average. Thus, in other words, China is a country of water shortage (Ding, 2011; Li & Li, 2012; Liu, 2013; Qian & He, 2011; Wang & Wang, 2012; Zhang, Li, & Tan, 2012).

(1) Uneven distribution of water resources

Water resources distribution in China is quite disproportionate, both spatially and temporally. For instance, the area of the watershed of the Yangtze River and the region south to the Yangtze River only accounts for 36.5% of China's total area, but with more than 81% of the total water resources in China. On the other hand, less than 20% of the total water resources are disproportionately distributed within the area north and west of the Yangtze River, where more than 70% of the total arable land in China is located. Furthermore, there are prominent conflicts between water supply and demand, especially in north, northwest and southwest China and in cities located in coastal regions of China, due to significant geographical differences in regional economic development and population distribution. Consequently, by the end of 20th century among 650 cities (with population over 2.00×10^5) in China, water shortage is one of the common obstacles to be overcome for more than 400 cities, severely hindering the overall development of China. On an average, the deficit of water shortage is near 5.0×10^{10} m^3 annually.

(2) Surface and ground water both heavily polluted

Generally, the water bodies of surface and ground water in China are polluted. With the increasing pollution from point and non-point sources, the situation of water pollution is deteriorating. The complexity, regionality, and permanency of water pollution have become the most severe and prominent problems of water resources in China. For example, by the end of 2010, the attainment rate of water quality in water functional zones was only 46%. Among 667 drinking water sources for centralized supply (for more than 1000 people), the attainment rate of the year-round water quality (at least 12 samples per year) was only 53%; and there were 37 drinking water sources, where the year-round water quality was completely non-attaining, accounting for 5.5%. Moreover, among 763 water quality monitoring stations, in about 62% of the stations where the water quality was categorized as Class-IV or Class-V, according to the Standards of Surface Water Quality (GB 3838—2002), as shown in Box 18.1.

Box 18.1 The classifications of surface water quality

Please be noted that the classifications of surface water quality can be defined as the following:
- *Class-I: sources of water, and national natural reserves;*
- *Class-II: first class of drinking water sources for centralized supply, habitats for valuable and rare aquatic organisms, spawning sites for fish and shrimp, feeding sites for juvenile fish, etc;*
- *Class-III: second class of drinking water sources for centralized supply, wintering sites for fish and shrimp, aquaculture farms of aquatic products, swimming passages and areas, etc;*
- *Class-IV: for industrial applications and not directly human contacted recreational purposes; and*
- *Class-V: for agricultural application and other scenic purposed.*
Please refer to GB 3838—2002 for more detailed information.

---

(3) Ecological degradation of water environment

With the rapid increase in demand for water due to booming economic growth, social development, and land exploitation and utilization, some serious problems related to the ecological degradation of the water environment have gradually emerged, such as cutoff of rivers, shrinkage of lakes, decrease in wetlands, reduction of biodiversity, degradation of the biotope, and so on. Moreover, the functions of freshwater systems have been severely degraded, especially for north China owing to over-extraction of groundwater. As estimated, the over-exploitation of groundwater is more than 2.0×10^{10} m^3, annually, resulting in more than 160 groundwater over-exploitation zones covering an area of 1.90×10^5 km^2. Consequently, land subsidence, saltwater intrusion, and other geological environment problems have inevitably ensued.

(4) Frequently occurred extreme weather events (floods and droughts)

Thanks to the significant increase in the water cycle rate on account of global warming, extreme weather events such as droughts, floods, intense rainfall, and typhoons have frequently occurred in China. In recent years, there has been a considerable increase in the intensity of droughts and floods. In addition, the drought areas within major agricultural zones in north China have been gradually expanding, and the frequency of extreme droughts and floods has drastically increased, for instance, heavy flooding of the Huaihe River in 2003 and 2007, and of the Zhujiang River in 2005 as well as severe droughts in Sichuan and Chongqing in 2006, in Xinjiang in 2008, in north China in 2009 and 2011, and in southwest China in 2010. According to statistics, from 1991 to 2010, nine big droughts occurred in China. Additionally, the occurrence of anthropogenic water pollution events and malfunctions of urban water supply system have shown a substantial increase.

18.2 Water Pollution in China

In China, along with rapid industrialization and urbanization, water pollution has posed a serious threat to the ecological environment, including surface water, ground water, soil, offshore sea, and even the atmosphere. In addition, the safety of water resources was seriously jeopardized to significantly compromise the safety of drinking water and agricultural products, and eventually impaired human health. Though the overall situation of water pollution has witnessed a gradual improvement, there is still much work to be done (Yu, 2014; Zhang, 2014).

(1) Surface water pollution

Surface water is polluted mainly by domestic and industrial wastewater. According to the Bulletin of China Environmental Status Quo of 2016, the distribution of water quality classifications for 1,940 surface water quality monitoring stations were 2.4% (Class-I), 37.5% (Class-II), 27.9% (Class-III), 16.8% (Class-IV), and 6.9% (Class-V), based on GB 3838—2002. In particular, for those segments of rivers running through urban areas, surface water is mostly polluted by organic matter contained in both domestic and industrial wastewater, seriously jeopardize drinking water safety and public health.

(2) Watershed water pollution

Along with increasing agricultural development, industrial construction, and urban expansion, watershed pollution has gradually become a significant issue that hinders economic growth in China. According to the Bulletin of China Environmental Status Quo of 2016, the distribution of water quality classifications for 1,617 watershed water quality monitoring stations located at major rivers in China were 2.1% (Class-I), 41.8% (Class-II), 27.3% (Class-III), 13.4% (Class-IV), 6.3% (Class-V), and 5.1% (Worse than Class-V), based on GB 3838—2002. These rivers include the

Yangtze River, Yellow River, Pearl River, Songhuajiang River, Huai River, Hai River, and Liao River, and rivers in Zhejiang and Fujian Provinces, and in northwest and southwest China. The major indices exceeding the standards were chemical oxygen demand, COD (17.6%), total phosphorous, TP (15.1%), and biochemical oxygen demand, BOD5 (14.2%). The Hai River was severely polluted, while the Yellow River, Songhuajiang River, Huai River and Liao River were slightly polluted.

(3) Ground water pollution

There are four key types of ground water pollution in China: ① saltwater intrusion due to excessive exploitation of fresh ground water in coastal regions; ② nitrate pollution resulting from industrial and agricultural wastewater effluents; ③ contamination by petroleum and petrochemical products; ④ pollution caused by leachate from garbage landfill sites. According to the Bulletin of China Environmental Status Quo of 2016, the distribution of water quality classifications for 6,124 ground water quality monitoring stations were 10.1% (Class-I), 25.4% (Class-II), 4.4% (Class-III), 45.4% (Class-IV), and 14.7% (Class-V), according to the Standards of Ground Water Quality (GB/T 14848—2017 & DZ/T 0290—2015), as shown in Box 18.2. The key indicators of ground water quality exceeding the standards were manganese, iron, total hardness, total dissolved solids, nitrates, nitrites, ammonia nitrogen, sulfates, fluorides, etc., and in some ground water monitoring stations, heavy metals (such as arsenic, lead, cadmium, and hexavalent chromium).

Box 18.2 The classifications of ground water quality
Please be noted that the classifications of ground water quality can be defined as the following:
- *Class-I: low background concentrations of natural chemical compounds, suitable for all applications;*
- *Class-II: medium background concentrations of natural chemical compounds, suitable for all applications;*
- *Class-III: according to the reference value of human health, suitable for centralized drinking water supply, and for industrial and agricultural applications;*
- *Class-IV: according to the demands of agricultural and industrial usage, suitable for agricultural and partial industrial applications, and for drinking water supply after proper process and treatment; and*
- *Class-V: not suitable for drinking water sources, and other usages with care and discretion.*

Please refer to DZ/T 0290—2015 for more detailed information.

(4) Urban water pollution

Along with booming industrialization and rapid urbanization, domestic sewage and industrial wastewater have become two major causes of urban water pollution, due to a highly concentrated urban population and existence of various industrial complexes. In addition, agricultural, industrial and municipal wastes also play an important role in urban water pollution. According to the Bulletin of China Environmental Status Quo of 2016, among 1,235 water quality monitoring stations of drinking water sources for centralized supply (338 at county level and 897 at city level), there was 811 stations with complete attainment of the year-round water quality accounting for 90.4%. For 897 water quality monitoring stations at the city level, there were 563 surface water quality monitoring stations and 334 ground water quality monitoring stations. There were 527 surface water quality monitoring stations with complete attainment of the year-round water quality accounting for 93.6%, where the major pollutants exceeding standards were total phosphorous, sulfates, and manganese. Besides, there were 284 ground water quality monitoring stations with complete attainment of the year-round water quality accounting for 85.0%, where the major pollutants exceeding standards were manganese, iron, and ammonia nitrogen.

18.3 Laws and Regulations on Water Pollution Prevention and Control

The establishment of the legal system of water pollution prevention and control in China was initiated in the mid-1950s, with various regulations, standards, and guiding normative documents published, such as the *Standards of Drinking Water Quality* (December 1956), the *Regulations of Domestic Drinking Water Hygiene* (November 1959) and the *Standards of Domestic Drinking Water Hygiene (on Trial)* (December 1976). However, it was not until the enactment of the EP Law (on Trial) in 1979 that water pollution prevention and control was regulated in a statutory form, for the first time, with some principles stipulated.

Since the early 1980s, in attempts to cope with the deteriorating situation of the water environment, many laws and regulations of water pollution prevention and control have been publicized to gradually establish the fundamental legal system of water pollution prevention and control. For example, water environment protection was described in the Constitution (1982), and the measures of water pollution prevention and control were depicted in the EP Law (1989). The first integrated special regulation of water pollution prevention and control, the Water Pollution Law, was enacted in May 1984, and the *Implementation Provisions for the Law of the People's Republic of China on Water Pollution Prevention and Control* was publicized in July 1989. These two regulations greatly promoted the development of water pollution prevention and control, mitigated the situation of water pollution, and enhanced the institution of legal system for water pollution prevention and control. Later on, a series of emission standards of water pollutants were announced, as shown in Box 18.3.

> **Box 18.3 Some emission standards of water pollutants**
>
> - *the Environmental Quality Standards for Surface Water (GB3838, initially enacted in 1983, amended in 1988, 1999 and 2002).*
> - *the Comprehensive Wastewater Emission Standards (GB8978, initially enacted in 1988, revised in 1996).*
> - *the Regulations of Pollution Prevention and Control Management for the Protecting Zones of Drinking Water Sources (initially published in 1989, revised in 2010).*
> - *the Standards of Ground Water Quality (GB/T14848, initially enacted in 1993, revised in 2017).*
> - *the Emission Standards of Water Pollutants from Medical Institutions (GB18466, initially enacted in 2005).*
> - *the Emission Standards of Water Pollutants from Saponin Industry (GB20425, initially enacted in 2006).*
> - *the Emission Standards of Pollutants from Coal Industry (GB20426, initially enacted in 2006).*
> - *the Ordinance of Urban Drainage and Sewage Treatment (enacted in 2013).*

In addition, to counteract the extremely deteriorating condition of water bodies on account of water pollution and to preserve water resources, many regulations were promulgated. For example, in 1995, the *Interim Ordinances of Water Pollution Prevention and Control for Huaihe River Watershed*, the first special administrative regulation of water pollution prevention and control for a major river system, was enacted (revised in 2011) to denote the significant action of water pollution prevention and control for river systems. In 2004, the *Measures of Management for Supervision on Effluent Outlets to Rivers* was passed (revised in 2015) to protect water resources by enhancing the supervision and management of effluents into rivers. In 2006, the *Ordinances of Management for Water Drawing Permit and Water Resource Fee Collection* was decreed to promote the conservation and rational utilization of water resources. In 2011, the *Ordinances of Management for Taihu Watershed* was publicized to improve water resources protection and water pollution prevention and control in the Taihu watershed.

The Water Pollution Law (1984) is the first special and comprehensive law to focus on water pollution prevention and control, containing the principles, supervision, management and institution system of pollution prevention and control for surface and ground water. With rapid industrialization, urbanization, and economic growth, the situation of water pollution has radically changed, making the legal system outdated and inappropriate. To cope with constantly changing circumstances, the Water Pollution Law (2017) was enacted to include 103 articles within eight chapters, through various major revisions in 1996, 2008, and 2017. Furthermore, on April 2, 2015, the *Action Plan for Water Pollution Prevention and Control* (the Action Plan for Water Pollution or the Ten Tactics of Water Pollution) was officially disseminated by the State Council to clearly expound the detailed procedures of ten tactics (35 articles) of water pollution prevention and control, as shown in Box 18.4.

Box 18.4 Ten tactics of the Action Plan for Water Pollution Prevention and Control
- *Tactic-1: To control the pollutants emissions;*
- *Tactic-2: To promote the transformation and upgrade of economic structure;*
- *Tactic-3: To emphasis on water resources conservation and protection;*
- *Tactic-4: To strengthen technical support;*
- *Tactic-5: To facilitate the functions of market mechanism;*
- *Tactic-6: To intensify the execution and supervision of environmental laws;*
- *Tactic-7: To enhance the water environment management;*
- *Tactic-8: To ensure the environmental security of water ecology;*
- *Tactic-9: To explicitly elucidate the responsibility of each stakeholder;*
- *Tactic-10: To enhance the public participation and social supervision.*

18.4 Water Management in China

In order to rationally exploit, utilize, conserve, and protect water resources, the *Law of the People's Republic of China on Water* (the Water Law) was promulgated in March 1988 (revised in August 2002, August 2009, and July 2016) to conform to the needs of national economic and social development and to realize the sustainable utilization of water resources. Regarding water management in China, there is a huge water management system, containing various governmental departments with diverse responsibilities of water environment protection, including the former MEP (which became the MEE in March 2018), the Ministry of Water Resources (MWR), the Ministry of Housing and Urban-Rural Development (the MHURD), the former Ministry of Agriculture (which became the Ministry of Agriculture and Rural Affairs, the MARA, in March 2018), the former Ministry of Land and Resources (which became the Ministry of Natural Resources, the MNR, in March 2018), the former State Forestry Administration (SFA, which became the State Forestry and Grassland Administration, the SFGA, in March 2018), the Ministry of Transport (MOT), the Ministry of Industry and Information Technology (MIIT), the Ministry of Finance (MOF), the NDRC, and the former Ministry of Health, the MOH (which was integrated into the former National Health and Family Planning Commission, the NHFPC, in 2013, and reformed into the National Medical Security Bureau, the NMSC, in March 2018), as shown in Tables 18.1 and 18.2 (Chen, Gao, & Li, 2015; Lian & Wu, 2007; Kuang & Huang, 2013; Wang & Xue, 2009; Yu, 2014).

Table 18.1 Major responsibilities of water pollution prevention and control for various governmental agencies

Governmental agencies	Major responsibilities of water pollution prevention and control
The MEE	In charge of water pollution prevention and control; prepare the plan of water prevention and control plans for key areas and watersheds; instruct and coordinate the solutions for any major environmental pollution issues within locals, various departments, transboundary and cross-watersheds; investigate and manage the events of severe environmental pollution and ecological destruction; mediate trans-provincial pollution disputes; organize and coordinate water pollution prevention and control works for key watersheds; in charge of environmental supervision and auditing for environmental protection administration; organize the inspection on environmental protection enforcement; and publish regularly the status quo of environment quality of major cities and watersheds
The MWR	Collaborate with the competent authorities to supervise and manage water pollution prevention and control work; organize the division of water functional zones and control the effluents within the drinking water source areas; monitor the water resources and water quality of rivers, lakes and reservoirs; determine the regional water environmental capacity; propose the measures for total emission cap; coordinate and arbitrate any disputes of water affairs between departments and provinces; and publicize the National Bulletin of Water Resources
The MHURD	Planning, construction and management for the engineering of urban and industrial water conservation, urban water supply and sewage, and wastewater treatment
The MARA	In charge of pollution control for area sources, water environment protection for fishery, and habitat protection for wild aquatic animals
The MNR	In charge of marine environmental protection
The SFGA	In charge of the management for watershed ecology, wetland and water conservation forest protection
The MOT	In charge of the management for water transport environment, pollution control of water transport channels
The MIIT	Initiate supporting policies for water pollution prevention and control industries; draw up the policies and regulations for cleaner production industries related to water pollution prevention and control
The MOF	Participate in policies institution and funds management for pollution fees; participate in policies institution for wastewater treatment fees
The NDRC	In charge of planning for water resources exploitation and development and ecological and environmental construction, and the management for the balance between agricultural, forestry and hydrological development plans and policies
The MOH	In charge of supervision and management for drinking water source standards

Table 18.2 Major responsibilities of water resources management for various governmental agencies

Governmental agencies	Major responsibilities of water resources management
Environmental protection	Participate in instituting policies for water resources protection and related; participate in compiling plans for water resources protection; review the Environmental Impact Assessment Statements of water resources engineering projects
Water resources	In charge of water resources management; draw up plans for water resources protection; monitor the quantity and quality of various water bodies; publicize the Bulletin of National Water Resources; organize and implement the Permit System of Water Consumption and the Taxation System of Water Resources Fee; organize and manage major water resources engineering projects; initiate the policies and plans for water conservation and related standards; organize, instruct and supervise the implementation of water conservation
Housing and urban-rural development	In charge of drinking water management, urban water supply, urban water conservation management, and urban water affairs management
Agriculture and rural affairs	In charge of agricultural water sources protection, agricultural water consumption management, agricultural water consumption and agricultural irrigation conservation
Natural resources	In charge of ground water resources management and mineral water exploitation and management
Forestry and grassland	In charge of the protection of water conservation forest
Industrial information and technology	In charge of the management for industrial water consumption; draw up the standards of industrial water quota; the management for industrial water conservation; participate in the drafting of the standards of water resource fee and policies of water price
Finance	Participate in policies institution for water price, water resources fees and subsidy for people having difficulties to access water
NDRC	Organize, draw up and adjust the policies for water price; the infrastructure construction of water conservation and water resources

The water management system in China can be stratified as working on two levels: the central level and provincial level. At the central level, the MWR of the State Council is the competent authority for water management in charge of water resources planning and management, and supervising the pivotal water conservancy projects and the work of flood control, irrigation, water supply, and agricultural water conservancy, nationwide. Under the MWR, several committees of watershed management are established for seven major rivers (the Yangtze River, Yellow River, Pearl River, Liaohe River, Songhuajiang River, Haihe River and Huaihe River) and the Taihu. At the provincial level, bureaus of water resources are the competent authorities for water management.

In addition, led by the Vice Premier as the General Commander, the Minister of the MWR, the Vice Director of the NDRC and the Deputy Secretary General of the State Council as the Vice General Commanders, the State Flood Control and Drought Relief Headquarter (the SFCDRH) was established under the State Council as an ad hoc organization, in 1992, to take charge of commanding the work of flood control and drought relief. In addition, the General Office of State Flood Control and Drought Relief Headquarters is founded under the MWR to manage the daily routines of this organization, including the unified dispatch of flood control and drought relief for major rivers and key water conservancy projects, and the conservation of water and soil, nationwide.

The system of water pollution prevention and control in China can be characterized as the integration of centralized management and regionalized and departmentalized implementation. The former SEPA of the State Council, as the competent authority for environmental protection, was in charge of centralized management for water pollution prevention and control, including organizing the preparation and supervising the implementation of water pollution prevention and control plans for designated key areas and major watersheds; summoning the compilation of environmental functions zoning; drafting and enacting laws, regulations, provisions, and rules of water pollution prevention and control; directing, and coordinating solutions for local, regional, departmental and cross-basin major environmental issues; investigating and coping with significant environmental pollution incidents; arbitrating disputes concerning trans-provincial environmental pollution; organizing and coordinating the water pollution prevention and control for major watersheds; executing environmental supervision and auditing the environmental protection administration; and promulgating the status quo of water environmental quality of major cities and watersheds, regularly.

At the local level, the competent authorities for environmental protection of local governments, for example, the bureaus of environmental protection, are in charge of managing and supervising the implementation of water pollution prevention and control, locally. In addition, the competent authorities for water resources of local governments, for example, the bureaus of water resources will collaborate and coordinate with the bureaus of environmental protection to supervise the implementation of water pollution prevention and control. Furthermore, the main responsibilities of the bureaus of water resources are to prepare the plans for water resources protection, propose water functional zones, control the effluents discharged into drinking water sources and catchments, monitor the water yield and water quality of reservoirs and rivers, evaluate the carrying capacity of water bodies and to determine the emission caps, coordinate and arbitrate the disputes of water affairs between departments and provinces, and issue the Bulletin of National Water Resources, regularly.

18.5 Progress

Through continuous efforts in water governance, the water environment has been gradually improved by the work in the field of water pollution prevention and control.

(1) Continuing improvement of the legal system of water pollution prevention and control

The Water Pollution Law (1984) is the first special law on water pollution to set up the fundamental legal institution of water pollution prevention and control in China, including pollutants emission limits, pollution emission charges, application for emission, legal liability, environmental standards system, and so on. For more than 30 years, the Water Pollution Law has been revised several times (1996, 2008 and 2017) in response to the constantly changing status of water pollution, effectively and in a timely manner, together with other laws, regulations, ordinances, standards, etc., as shown in Table 18.3.

(2) Establishment of the supervision and management system on water environment with diverse government departments and agencies

The are many government departments and agencies involved in water governance, including the MEE, MWR, MHURD, MARA, MNR, SFGA, MOT, MIIT, MOF, NDRC and NOH, as described in Sect. 18.4. Please refer to Boxes 18.1 and 18.2 for more detailed information.

(3) Gradual perfection of supervision and management instruments and tools for water pollution prevention and control

The command-and-control type of supervision and management instruments and tools (emission standards, total emission cap, emission permit, etc.) has been continuously improved to play a basic and essential role in water pollution prevention and control. Meanwhile, diverse instruments and tools, including green tariff, environmental tax, ecological compensation, emission trade, green finance, green insurance, etc., have been progressively implemented through certain pilot programs to gradually form a framework of economic types of supervision and management instruments and tools.

(4) Institution of an accountability mechanism and performance evaluation system for water pollution prevention and control

To comprehensively evaluate the performance of leading groups and leading cadres, a liability and performance evaluation system on water pollution prevention and control for key watersheds and water conservancy projects has been instituted to promote the implementation of water pollution prevention and control for key watersheds and to fulfill the objective liability system. The strictest water resources management and

Table 18.3 Major shifting characteristics of water governance strategies in the Water Pollution Law

Year	Key features of the Water Pollution Law
The Water Pollution Law (1984)	Set up the fundamental legal institution of water pollution prevention and control in China, including pollutants emission limits, pollution emission charges, application for emission, legal liability, environmental standards system, and so on
The Water Pollution Law (1996)	From point sources control to area sources, watershed and regional integrated governance and control From end-of-pipe control to sources control and comprehensive industrial production control From pollutant concentration control to total emission cap control From decentralized point sources control to the combination of centralized and decentralized control
The Water Pollution Law (2008)	To intensify the responsibility and liability of local governments in water pollution prevention and control To improve the supervision and management system of water pollution prevention and control To expand the work scope of water pollution prevention and control To focus on the protection of potable water sources To enhance the law enforcement jurisdiction of the competent authorities for environmental protection To elevate the criminal action and penalty against environmental violations
The Water Pollution Law (2017)	To further intensify the responsibility and liability of local governments in water pollution prevention and control To delimit the boundary of environmental violations and no exceeding of total emission cap To further enhance the total emission cap control for key water pollutants To comprehensively implement the emission permit system and to regulate the emission of enterprises To improve the water environmental monitoring network to establish the unified publicizing mechanism of water environmental information To initiate the River Leader System in the Water Pollution Law To strengthen the rural and agricultural wastewater treatment

performance evaluation system has been initiated to explicitly define the organization, procedures, content, assessment, and reporting of the performance evaluation system. The river leader system, the unique performance evaluation system, has been implemented by appointing the governors of local governments as the river leaders in charge of the water pollution prevention and control work for rivers and watersheds.

(5) Elevation of capacity building of supervision and management for water pollution prevention and control

Since its institution in 1981, the environment statistics system has been steadily enhanced through several modifications in scope and methods. The water environment monitoring network has been preliminarily founded through the collaboration of multiple governmental departments and agencies. Supervising the monitoring of pollution sources has been fully implemented, with support from special funds for emission reduction, to develop an emission reduction monitoring system as part of a major pollutants emission monitoring system. The performance evaluation system for the emission reduction of total emission cap is being constantly improved through the promulgation of diverse supporting measures and schemes.

References

Chen, J., Gao, S., & Li, Z. (2015). Status, problems and improvement for the supervision system of water environment in China. *Development Research, 2,* 4–9.

Ding, W. (2011). Strategy and recommendations on sustainable development of water resources in China. *Chinese Agricultural Science Bulletin, 27*(14), 221–226.

Kuang, Y., & Huang, N. (2013). Several issues about the research on the water resources utilization and water environment protection in China. *China Population, Resources and Environment, 23*(4), 29–33.

Li, J., & Li, L. (2012). Water resources supporting capacity to regional socio-economic development of China. *Acta Geographica Sinica, 67*(3), 410–419.

Lian, Y., & Wu, J. (2007). *Chinese national conditions report.* Beijing: China Economic Publishing House.

Liu, N. (2013). On the coordinated routine and emergency management of China's hydrology and water resources. *Advances in Water Science, 24*(2), 280–286.

Qian, W., & He, C. (2011). China's regional difference of water resource use efficiency and influencing factors. *China Population, Resources and Environment, 21*(2), 54–60.

Wang, H., & Wang, J. (2012). Sustainable utilization of China's water resources. *Bulletin of Chinese Academy of Sciences, 27*(3), 352–359.

Wang, H., & Xue, H. (2009). Problems and countermeasures of the work to prevent and control water pollution in China. *Environmental Science and Management, 34*(2), 24–27.

Yu, B. (2014). *A comparative study on water resource pollution prevention and treatment of China and America.* Master: China University of Geosciences (Beijing).

Zhang, J., Li, J., & Tan, Y. (2012). Analysis of the spatio-temporal matching of water resource and economic development factors in China. *Resources Science, 34*(8), 1546–1555.

Zhang, X. (2014). Trend of and the governance system for water pollution in China. *China Soft Science, 10,* 11–14.

Chapter 19
Land Resources Governance in China

19.1 Land Resources in China

China is the third-largest country in the world, after Russia and Canada. China's territory possesses several unique characteristics (Liu, Xia, & Song, 1999; Ma, 2005; MNR, 2018).

(1) Vast land with diverse types of land cover

Within China's vast land area, there are diverse climate patterns ranging from cold zone, temperate zone, to tropical zone, including temperate zone (25.9%), warm temperate zone (18.5%) and subtropical zone (26.0%). Consequently, various land covers, wet area (32.2%), semi-wet area (17.8%), semi-dry area (19.2%) and dry area (30.8%), are allocated on China's territory to develop into many different forms of land resources to effectively facilitate the overall development of agriculture, forestry, animal husbandry, and fishery.

(2) Large portion of mountainous and hilly areas, small portion of flat and plain areas

China is a very mountainous country with many mountains, plateaus, and hills (69%) and a few plains (31%). Generally, the features of mountain areas are high altitude drop, steep slope, thin soil layer, and not suitable for tillage, greatly restricting agricultural development. However, the mountain area in south China is very suitable for forestry development due to high temperature and moisture. The mountain area in north and northwest China, mainly suitable for animal husbandry, is the catchment area of the agricultural irrigation water source for the plain area, which is an essential part of the natural resources for significantly promoting regional development.

(3) Small agricultural land per capita with massive agricultural land

According to the *2017 Statistical Bulletin of Chinese Land, Mineral and Marine Resources*, by the end of 2016, the total area of agricultural land is 6.451266×10^6

© Chemical Industry Press and Springer Nature Singapore Pte Ltd. 2020

km^2, including arable land (1.349210 × 10^6 km^2), garden land (1.42663 × 10^5 km^2), forest land (2.529081 × 10^6 km^2), and grassland (2.193592 × 10^6 km^2); the total area of construction land is 3.90951 × 10^5 km^2, including cities, villages, towns and industrial sites (3.17947 × 10^5 km^2). Apparently, the areas of various land covers are large, but the land per capita is quite small due to enormous population (1.38 billion). For example, the land per capita in China is about 0.007 km^2, less than one-third of the world average; the arable land per capita in China is about 0.001 km^2, which is far less than that in Canada (0.016 km^2), the United States (0.008 km^2) and India (0.0017 km^2) (MNR, 2018).

(4) Insufficient reserve land resources

According to the statistics from the State Forestry and Grassland Administration (the SFGA), the reserve land resources for future development in China is only 1.225 × 10^6 km^2, including sparse trees (1.56 × 10^5 km^2) and scrublands (2.96 × 10^5 km^2). Among these, the area of barren mountains and lands that can be suitable for forest and grazing is about 9.00 × 10^5 km^2. Further, the area suitable for cultivation and artificial pasture is around 3.3 × 10^5 km^2, where the area suitable for grains and cotton crops is near 1.3 × 10^5 km^2. On the contrary, the area that is difficult for humans to exploit and utilize, such as marshlands, moving dunes, deserts, the Gobi Desert, lands over 3000 m in altitude, and so on, is close to 3.487 × 10^6 km^2, accounting for 36.3% of the entire territory.

(5) Extreme imbalanced distribution of land resources with huge difference in land productivity

As depicted earlier, there are diverse classifications of land cover allocated within various climate zones in China. Noticeably, the area located in the monsoon area in southeast and south China is an important agricultural and forest zone with higher land productivity, due to its warm climate and sufficient water supply. However, in addition to frequently occurring natural disasters, this is a hilly area causing agricultural land fragmentation, which has seriously limited the integrated agricultural development. In contrast, the area in northwest and north China is that of grassland, desert steppe, and arid desert. Although due to an abundance of sunshine and heat, the annual precipitation is less than 250 mm, hindering agricultural development. In southwest China, the average altitude is over 3000 m. Although it has sufficient sunshine, the lack of heat has become the greatest disadvantage for agricultural development. Overall, the land resources distribution in China is quite imbalanced with significant regional differences.

19.2 Land Resources Protection

Land resources are the foundation of social and economic development. Land resources protection can be primarily categorized into the following three aspects.

(1) Protecting the quantity of land resources

Along with rapid population and economic growth, the quantity of land resources will play a critical role in affecting social and economic development. The quantity of land resources is defined as the total area of horizontal land resources, including diverse types of land resources and the area of each land resource type. In other words, protecting the quantity of land resources mainly refers to preservation of land resources, such as protecting agricultural land and preventing excessive expansion of non-agricultural land, through fundamental farmland protection.

(2) Protecting the quality of land resources

The quality of land resources is determined through land resources evaluation. Land resources evaluation means to appreciate the value of land resources for certain types of utilization. Therefore, the quality of land resources can be determined by the degree of appropriateness, the potential of productivity, the worth and the pollution status quo of land resources. To protect the quality of land resources, especially for cultivated land resources, means to protect the soil fertility of land resources, to maintain the potential of productivity of land resources, and to enhance the production levels of land resources.

(3) Protecting the ecological environment of land resources

Good ecological environment of land resources is one of the essential prerequisites for various uses of land. The balance of ecological environment of land resources is the primary index for evaluating the environmental benefit of land resources protection. To maintain the balance of ecological environment of land resources is the primary work of land resources protection, including establishing and maintaining the ecological balance of land resources and ensuring the sustainable utilization of land resources. Thus, the ecological environment protection of land resources is mainly composed of pollution prevention and control, maintenance of grassland vegetation, prevention of desertification, salinization, and water and soil erosion, and water resources protection of land resources.

19.3 Land Resources Degradation and Land Pollution in China

Triggered by various factors, including natural processes (such as environmental change, ecological evolution, and natural disasters) and anthropogenic causes (such as artificial disruption and destruction, and irrational exploitation and development), land degradation has seriously restricted agricultural development. There are several major types of land degradation, including erosion, desertification, salinization, gleying process, and pollution (MWR and NBS, 2013; SFA, 2015; MEP, 2017; MEE, 2018).

19.3.1 Land Degradation

According to the latest report, the *Bulletin of the First National Census for Water*, the total area of soil erosion in China is 2.949×10^6 km^2, where hydro-erosion is 1.293×10^6 km^2 and wind-erosion is 1.656×10^6 km^2, by the end of 2012 (MWR and NBS, 2013). Based on the *Bulletin of the Fifth National Desertification and Sandification Land Monitoring*, in China, the area of desertification is 2.6116×10^6 km^2 and the area of sandification is 1.7212×10^6 km^2, by the end of 2014 (SFA, 2015). Compared with the statistics of 2009, the area of desertification has reduced by 1.212×10^4 km^2 and the area of sandification has declined by 9.902×10^3 km^2.

19.3.2 Land Pollution

According to the *Bulletin of the First Nationwide Survey on Soil Contamination Status Quo* (from April 2005 to December 2013), about 16.1% of total sampling points across surveyed land (covering near 6.3 million km^2) is nonattainment, to various extents, including slight (11.2%), mild (2.3%), moderate (1.5%), and severe (1.1%), where there are various kinds of pollution, including inorganic pollution (82.8%), organic pollution, radioactive pollution, and biological pollution (Song, Chen, & Liu, 2013; MEP and MLR, 2014; Zhuang, 2015; Zang & Li, 2018).

(1) Inorganic pollution

Major inorganic pollutants are cadmium, mercury, arsenic, lead, chromium, copper, zinc, selenium, fluorine, and so on, resulting from chemical fertilizers, wastewater irrigation and industrial wastes. In China, about $(2.0-2.5) \times 10^5$ km^2 of cultivated land is contaminated by heavy metals, such as cadmium, arsenic, lead, and chromium, where 1.0×10^5 km^2 is polluted by industrial wastes and 3.3×10^4 km^2 is polluted by wastewater irrigation.

(2) Organic pollution

Organic pollution is mainly caused by organic pesticides, phenols, synthetic detergents, and so on, resulted from overusing pesticides, residues of industrial wastes, municipal garbage, and so forth. For example, about $(8.6–10.7) \times 10^3$ km^2 agricultural land is polluted by pesticides at different levels in China.

19.3.3 Overall Spatial Distribution of Soil Pollution

In general, the situation of soil pollution in south China is worse than that in north China, with some of most prominent areas being the Yangtze River Delta Economic Zone, the Pearl River Delta Economic Zone, and the Old Heavy Industry Base in Northeast China. In southwest and mid-south China, soil contamination is mostly caused by heavy metals. The state of inorganic pollution by cadmium, mercury, arsenic and lead has generally worsened from the northwest to the southeast and from the northeast to the southwest.

19.3.4 Status Quo of Soil Pollution of Various Types of Land

According to the *Bulletin of the First Nationwide Survey on Soil Contamination Status Quo*, more than 19.4% of all sampling points for cultivated land is nonattainment, mostly contaminated by cadmium, nickel, copper, arsenic, mercury, lead, dichlorodiphenyltrichloroethane (DDT), polycyclic aromatic hydrocarbon (PAH), etc.; about 10.0% of all sampling points for forest land is nonattainment, mainly polluted by arsenic, cadmium, hexachlorocyclohexane (HCH), and DDT; nearly 10.4% of all sampling points for grassland is nonattainment, primarily tainted by nickel, cadmium, and arsenic; over 11.4% of all sampling points for undeveloped and unused land is nonattainment, largely poisoned by nickel and cadmium.

Moreover, the situation of industrial and mining sites is even worse. For example, around 36.3% of 5846 sampling points for 690 industrial sites (including ferrous metals, nonferrous metals, leather ware, papermaking, petroleum, coal, chemical engineering, pharmaceutical, chemical fiber, rubber and plastic, mineral product, metalware, electric power, and so forth) and their surroundings is nonattainment. Nearly 34.9% of 775 sampling points for 81 blocks of abandoned industrial wasteland (such as chemical engineering, mining, metallurgical industries) is nonattainment, generally polluted by zinc, mercury, lead, chromium, arsenic, and PAH. About 29.4% of 2523 sampling points for 146 industrial parks is nonattainment, majorly contaminated by cadmium, lead, copper, arsenic, and zinc (for metal smelting industrial parks and their surroundings), and PAH (for chemical engineering industrial parks and their surroundings). Among 188 solid waste disposal sites, 21.3% of 1351 sampling points is nonattainment, mainly polluted by inorganic contaminants. In addition, incinerating and landfill sites are mostly polluted by organic pollutants. For 13 oil extraction sites, 23.6% of 1351 sampling points is contaminated by petroleum hydrocarbon and PAH. For 70 mining sites and their surroundings, 33.4% of 1672 sampling points is polluted by cadmium, lead, arsenic, and PAH. Among 55 wastewater irrigation areas, 29 are nonattainment where 26.4% of 1378 sampling points is polluted by cadmium, arsenic, and PAH. About 20.3% of 1578 sampling points for 267 major highways (within 150 m along the highway) is nonattainment, mainly polluted by lead, zinc, arsenic, and PAH.

19.4 Laws and Regulations on Land Pollution Prevention and Control

The seriousness of land pollution is China has been gradually recognized since 2006. In countering the increasing significance of land pollution, many relevant governmental agencies, both central and local, have progressively publicized many policies and countermeasures to enhance the work of land pollution governance and remediation, as shown in Table 19.1. However, no systematic guiding framework has been

Table 19.1 Major responsibilities of soil pollution prevention and control for various governmental agencies

Date	Issued by	Name	Key features
2008.06	The MEP	Opinion Concerning to Strengthen the Work of Soil Pollution Prevention and Control	To establish the preliminary planning for soil environment pollution prevention and control by 2010, to establish the basic supervision and management system for soil pollution prevention and control by 2015
2009.12	The MEP	Technical Guidelines for Soil Remediation of Contaminated Sites (Exposure Draft)	To detail the content, the technologies and the engineering scheme of remediation work for contaminated sites
2011.03	The MLR	The Ordinance of Land Reclamation	To emphasize the importance of soil quality and eco-environment protection during the process of land reclamation
2012.11	The MEP	Notification Concerning the Environmental Safety for Re-development and Re-utilization of Industries and Enterprises Sites	To propose the fundamental management for industries and enterprises contaminated sites; to comprehensively investigate contaminated sites; to determine the liability, according to the "polluter pays principle"

(continued)

19.4 Laws and Regulations on Land Pollution Prevention and Control

Table 19.1 (continued)

Date	Issued by	Name	Key features
2013.01	The State Council	Recent Arrangements on Soil Environment Protection and Comprehensive Governance	To thoroughly survey nationwide soil status quo by 2015; to ensure the attaining rate of sampling sites of cultivated land is greater than 80%; to build the National Soil Environmental Protection System by 2020
2013.03	The MLR	Implementation Measures for the Ordinance of Land Reclamation	To set up the Special Funds for geological environment remediation and governance of mines
2014.02	The MEP	Technical Guidelines for Soil Remediation of Contaminated Sites	To provide technical guidance and support to environment status quo investigation, risk assessment and remediation governance
2014.04	The MEP	The 2014 Amendment of the EP Law	To include the content of soil remediation; to enroll the processing of the Soil Pollution Law as the first priority; to pass the Action Plan for Soil Pollution
2014.05	The MEP	Notification Concerning to Strengthen the Pollution Prevention and Control Work During the Closing, Shut-Down, and Relocation, and Re-development and Re-utilization On-Site	To deploy site environment survey for closed, halted and relocated industries and enterprises; to strictly control title transfer and re-development construction review of contaminated sites
2014.11	The MEP	Operating Instruction on Environment Investigation, Assessment and Remediation of Industries and Enterprises Sites (on Trial)	To publicize the status quo of soil and groundwater environment of industries and enterprises sites in time; to investigate the site environment of closed, halted and relocated industries and enterprises; to restrict title transfer and control re-development construction review of contaminated sites; to strengthen site investigation assessment and remediation monitoring
2016.02	The MEP	Guidelines for Soil Pollution Risk Assessment of Construction Land (Exposure Draft)	To regulate the initiation of risk screening and risk assessment of 118 species of soil pollutants

(continued)

Table 19.1 (continued)

Date	Issued by	Name	Key features
2016.05	The State Council	The Action Plan for Soil Pollution Prevention and Control	To contain the deteriorating trend of land pollution by 2020; to effectively protect agricultural land; to safeguard construction land; to achieve significant result in soil pollution prevention and control; to promote the cooperative mode of governmental and social capital in order to drive more social capital investment in soil pollution prevention and control
2016.11	The MEP, the MST	Outlines of Science and Technology Development Planning for the 13th Five-Year-Plan of National Environmental Protection	Will invest 30 billion CNY to construct several Key Laboratories of National Environmental Protection and Centers of National Environmental Protection Engineering, where 3 billion CNY will be spent on soil and groundwater pollution prevention and control
2016.12	The MEP, the MOF, the MLR, the MOA and the NHFPC	The General Planning for the Survey of Nationwide Soil Pollution Status Quo	To narrow down the survey targets with systematic and explicit scope, based on existing information; to reveal the area and distribution of contaminated agricultural land and its effects on the quality of agricultural products by 2018; to comprehend the distribution and environmental risk of contaminated soil from key industries and enterprises sites by 2020
2016.12	The MEP	The Management Measures for Soil Environment of Contaminated Land (on Trial)	For contaminated land to be re-developed and re-utilized for residential area, business, schools, medical services, nursing homes and other public facilities, the land use right holder should deploy governance and remediation, once governance and remediation are affirmed to be necessary by risk assessment; to compile the proposal of governance and remediation engineering for contaminated land

(continued)

19.4 Laws and Regulations on Land Pollution Prevention and Control

Table 19.1 (continued)

Date	Issued by	Name	Key features
2017.02	The NDRC	National Land Renovation Plan (2016–2020)	To ensure 2.67×10^5 km^2 of high quality cultivated land and to push forward to 4.0×10^5 km^2, with a total investment of CNY 0.72–1.08 trillion yuan; the renovation rate of cultivated land should be over 60%, nationwide; to increase cultivated land by 1.33×10^4 km^2 and to recondition medium and low quality cultivated land by 133,000 km^2; to regularize village construction land by 4000 km^2; to transform and develop low efficiency urban land by 4000 km^2
2017.04	The MEP	The Developing Plan for National Environmental Protection Standards during the 13th FYP	According to revised soil environment quality standards, to institute the Technical Specification for Soil Environment Quality Assessment, and to study the technical specifications for soil environment investigation, risk assessment, risk control, and governance and remediation of polluted sites
2017.04	The MST, the MEP, the MHURD, the SFA and the CMA	Special Plan for the Environmental Science and Technology Innovation during the 13th FYP	To explore the soil pollution prevention and control, and remediation technologies for agricultural land; to explore soil remediation and safe re-development technologies for polluted industrial sites; to explore soil pollution control and remediation technologies for solid waste disposal sites; to explore soil pollution control and integrated remediation technologies for mining sites; to explore the early warning monitoring and risk management technologies of soil pollution
2017.07	The MEP	Soil and Sediment: Digestion of Total Metal Elements Microwave Assisted Acid Digestion Method (HJ 832—2017)	To improve the soil standards system by increasing the standards of soil environment monitoring methods to 64 and expanding the targeting chemicals to 280

(continued)

Table 19.1 (continued)

Date	Issued by	Name	Key features
2017.08	The MEP	Technical Guidelines for Environmental Impact Assessment: Soil Environment (Exposure Draft)	To regulate the general principles, the content, working routines, methods and requirements for soil environmental impact assessment
2017.09	The MEP	Soil Environment Management Measures for Agricultural Land (on Trial)	To strengthen soil environment protection, supervision and management for agricultural land; to protect soil environment of agricultural land; to control soil environment risk of agricultural land; to ensure quality safety of agricultural products
2017.10	The MIIT	Guiding Opinion Concerning to Expedite the Development of Environmental Protection Equipment Manufacturing Industry	To set forth the direction of future research on core technologies and key equipment for water pollution prevention and control facility, solid waste treatment and disposal facility, soil pollution remediation facility, and so on
2017.12	The MEP	Handbook of Technical Evaluation for the Effectiveness of Soil Pollution Governance and Remediation (on Trial)	To guide provincial (or municipal) governments to comprehensively evaluated the effectiveness of soil pollution governance and remediation performed by the third party entrusted by provincial (or municipal) governments
2018.01	The MST, the MLR, the MEP, the MHURD, the MOA	The Catalogue of Advanced Technologies and Equipment for Soil Pollution Prevention and Control	To enlist 15 advanced technologies and equipment, including ex situ washing technique, the integrated equipment of automatic mixing for soil and medicament, and so on
2018.08	The MEP	The Law of People's Republic of China on Soil Pollution Prevention and Control (the Soil Pollution Law)	To deploy soil pollution investigation so to reveal the status quo of soil pollution; to define the liability of pollution and to delineate the principles of prevention and control; to set the options of technical paths and the norms of technical standards; to regulate the relationship between government mechanisms and market mechanisms so to ensure the sustainable operation of pollution governance and prevention

formed until *the Action Plan for Soil Pollution Prevention and Control* (the Action Plan for Soil Pollution or the Ten Tactics of Soil Pollution, as shown in Box 19.1) was promulgated by the State Council in May 2016, which has greatly strengthened the supervision and management of soil pollution through legislation, promoted the coerciveness and development of the land governance market, and driven special funds in supporting the land governance market. In order to facilitate the smooth implementation of the Ten Tactics of Soil Pollution, the *General Planning for the Comprehensive Survey of Nationwide Soil Pollution Status Quo* (the General Planning) was jointly promulgated by the MEP, MOF, MLR, MOA and National Health and Family Planning Committee (NHFPC), in October 2016, to highlight the core work and set unified standards for survey, through the integration of various professional expertise from diverse governmental agencies and private sectors. The purpose of General Planning is to find out the degree and distribution of soil pollution of agricultural land and their impacts on agricultural products by the end of 2018, and to fully apprehend the distribution of soil pollution of key industries and enterprises and their environmental risks by the end of 2020. In the *Law of the People's Republic of China on Soil Pollution Prevention and Control* (the Soil Pollution Law) (August 31, 2018), the investigation of soil pollution status quo, the liability of soil pollution, measures of soil pollution prevention and control, technical standards, and applicability, means of financial supports, and so on, have been comprehensively planned and deployed.

Box 19.1 Ten tactics of the Action Plan for Soil Pollution Prevention and Control

- *Tactic-1: To deploy thorough investigation on soil pollution so as to fully comprehend the status quo of soil environment quality;*
- *Tactic-2: To promote the legislation of soil pollution prevention and control and to establish the sound regulations and standards system for soil pollution prevention and control;*
- *Tactic-3: To implement classification management for agricultural lands and to guarantee the environmental safety of agricultural production;*
- *Tactic-4: To implement admission management for construction lands and to prevent residential environmental hazards and risks;*
- *Tactic-5: To strengthen the protection of unpolluted soil and to strictly control the growth of polluted lands;*
- *Tactic-6: To reinforce the monitoring on pollution sources and to concretely realize the tasks of soil pollution prevention;*
- *Tactic-7: To deploy pollution governance and remediation and to improve regional soil environment quality;*
- *Tactic-8: To intensify the research and development for land protection and to stimulate the development of environmental protection industry;*
- *Tactic-9: To fully utilize the leading role of governmental mechanisms and to construct the system of soil environment governance;*

> • *Tactic-10: To enhance the performance evaluation and to stringently identify the accountable liability.*

19.5 Challenges and Countermeasures

19.5.1 Challenges

In the context of massive population, scarce land resources, explosive economic growth and increasing environmental concerns, there are many problems and challenges facing land protection in China as summarized below, along with the intensifying pressure of food security.

(1) Increasing pressure of soil environment protection

In order to maintain continuous rational social development and economic growth in China, rapid urbanization and industrialization will certainly be inevitable and will result in the persistent discharge of various pollutants. As mentioned earlier, more than 15% of total sampling points across surveyed land in China are currently considered as nonattainment areas due to inorganic, organic, radioactive, and pathogen contamination. With increasing exploitation and consumption of mineral resources and energy resources, regional land environments would be further polluted or deteriorated by various heavy metals and organic pollutants. Additionally, in order to guarantee food security, intensive and extensive applications of agricultural chemicals, including chemical fertilizers, pesticides, and agricultural films, have worsened the existing unoptimistic situation of land environment.

(2) Increasing complexity of soil environment problems

In addition to heavy metals and organic pollutants, radionuclides, pathogens, antibiotics, environmental endocrine disruptors, etc., have gradually intensified their impacts on land quality, complicating the deteriorating situation of soil pollution in China. The increasing diversity and complexity of soil pollution have greatly augmented the environmental risk of soil pollution and significantly increased the difficulty of soil pollution control, becoming serious threats to public health and to economic and soil development.

19.5.2 Countermeasures

In response to the unoptimistic situation of land resources degradation and land pollution, many countermeasures should be implemented, in accordance with the concepts of sustainable development and ecological civilization (Ma, 2005; Sun, 2015).

(1) To promote sustainable development of land utilization through effective management

In order to fulfill regional sustainable development, it is of great importance to implement scientific and rational land utilization through reasonable zoning plans for various land utilizations, such as strictly controlling the quantity of land resources, prohibiting the transfer of agricultural land for non-agricultural utilization, fully protecting the quality of cultivated land resources, and so on. In addition, in order to achieve ecological civilization, it is necessary to advocate the concept of land conservation and sustainable utilization of land resources to effectively reduce and prevent the losses of land resources due to excessive and irrational applications on land resources.

Moreover, the legal system of land resources protection has been gradually improved and perfected through the promulgation of diverse statutes, laws, and regulations for land resources protection. Land resources protection and utilization have been greatly promoted and significantly maximized through the facilitation and enhancement of vertical and lateral cooperation and collaboration on land resources protection and utilization between various governmental departments and agencies involved in land resources protection and utilization, due to explicitly clarified and defined responsibilities and jurisdictions for each of them.

(2) To optimize the modes of land utilizations and to change the patterns of economic growth

Traditionally, the distinct features of resource exploitation and utilization in China are high investment and low profit, which have already caused the rapid depletion of land resources and severe destruction of the ecological environment. In order to conserve land resources, improve land productivity, and enhance the efficiency and effectiveness of resource utilization, intensive patterns of land resource utilization have been broadly applied, nationwide. Generally, the intensity and benefits of land utilization for three industries can be ranked as tertiary industry greater than secondary industry greater than primary industry. Thus, to optimize the intensity and benefits of land utilization, economic structure reform has been inevitably deployed throughout the country by shifting from extensive land resources utilization to intensive land resources utilization.

Before the Economic Reform and Openness initiated in the late 1970s, there was a significant imbalance between the allocation of primary, secondary, and tertiary industries. Through the optimization and upgradation of the economic structure, the proportion of each industry has been considerably improved, as primary industries have declined while secondary and tertiary industries ascend. The secondary industry has become the key impetus of Chinese economic growth and the role of the tertiary industry has been significantly enhanced. Furthermore, the development within each industry has been significant changed as well. For primary industry, diverse policies and ordinances have been promulgated to actively and effectively promote the intensive patterns of cultivation and plantation. For secondary industry, through technology upgrading and layout optimization, the importance of high and new technology industry (with less land demand) has steadily increased whilst heavy industry (with high pollution, high energy consumption, and high land demand) is still the core of secondary industry. For tertiary industry, the portion of the tertiary industry has gradually surpassed that of the primary industry, through the implementation of foresighted industry development strategy since the 7th FYP (1986–1990).

(3) To stabilize the quantity of cultivated land resources

Cultivated land resources have always been the essence of private property, based on traditional Chinese agricultural civilization. Owing to the large population, food security has constantly been the first priority of agricultural development in China. Faced with the rapid shrinkage of cultivated land resources since the late 1990s, many countermeasures have been promulgated to ensure the area of cultivated land. For example, in the 11th FYP (2006–2010), it was explicitly stipulated that the area of cultivated land should be no less than 1.2×10^6 km^2, known as the Redline of Cultivated Land. In addition, to elevate the flexibility of land resources utilization, a dynamic balancing mechanism, the ex situ compensation, was explicitly promulgated in the *Notice of Several Policy Measures Regarding the Enhancement of Cultivated Land Protection and the Promotion of Economic Development*, publicized by the former MLR in December 2000, where any encroachment of cultivated land by approved non-agricultural construction utilization should be compensated with land to be reclaimed as cultivated land at different site(s) or with the necessary expense to reclaim cultivated land at different jurisdiction(s), at an equivalent amount and quality. Furthermore, in order to facilitate the intensive cultivated land utilization, the transfer mechanism was clearly specified in the *Decision Concerning the Deepening Reform of Stringent Land Management*, published by the State Council in October 2004, where a farmer's Right of Land Use can be transferred to another farmer or an economic organization (with advanced professional skills, and bigger operation scale) through subcontracting, transferring the possession, becoming a shareholder, cooperating, leasing, and interchanging, so as to promote the effectiveness and efficiency of cultivated land resources.

(4) To protect the quality of land resources

Many schemes were implemented to protect the quality of land resources. In order to protect water sources, conserve water and soil, and prevent soil erosion, the farmland located in areas susceptible to wind and water erosion should be converted to forestland, grassland, or fallow field so as to protect the quality of land resources and maintain the productivity of land resources. For example, the Grains for Green

Program (the GGP), also known as Returning Farmland to Forestland, was initiated in 1999 with the intention of mitigating and preventing flooding and soil erosion, and includes two parts: reforestation on sloping cropland and afforestation on barren land. To ensure the concrete execution of the GGP, many economic schemes, including subsidies, allowances, compensation, financial aid, and various financial incentives, were applied, along with diverse preferential policies and technical support. Moreover, to protect land resources, enhance the efficiency of natural resources utilization, and equalize the benefit differences between various stakeholders, resources taxes were firstly imposed on coal, petroleum and natural gas for natural resources exploitation and extraction. In 1984, the *Ordinances for the Resources Taxes (Draft)* was originally publicized by the State Council. The *Interim Ordinances of Farmland Encroachment Tax* (April 1987) and the *Interim Ordinances of Urban Land Utilization Tax* (September 1988) were decreed in 1987 and 1988. The *Interim Ordinances of the Resources Taxes* was proclaimed in November 1993 to include ferrous metals minerals. Subsequently, non-ferrous metals minerals, non-metallic minerals and salts were included in the *Amendment of the Interim Ordinances of the Resources Taxes* on September 30, 2011. In the *Notice Concerning Promoting the Overall Reform on Resources Taxes*, promulgated on May 9, 2016 by the MOF and State Administration of Taxation (SAT), a resource tax was imposed on water resources (including both surface water and ground water) as a pilot program. In the future, forestland, grassland and wetland will be gradually subject to resources taxes.

In addition, more practical measures, such as the application of intensive land utilization, modifications to land resources with medium- and low-productivity, and reinforcement in cropland infrastructure, were implemented to prevent further deterioration of land quality and to enhance the productivity of land resources.

(5) To protect the ecological environment of land resources

Rapid industrialization and urbanization have resulted in severe environmental pollution and ecological degradation of land resources. Besides, for food security consideration, it is essential to ensure a stable food production. In such a context, excessive applications of chemical fertilizers and pesticides have seriously exacerbated the condition of soil contamination and water pollution. Faced with such severe challenges, many countermeasures have been implemented, as explicitly stipulated in the *Recent Work Arrangements for Soil Environment Protection and Comprehensive Governance*, promulgated by the State Council on January 23, 2013, including: ① strict control of any new soil pollution; ② identifying the area of soil environment protection with first priority; ③ strengthening the risk control of contaminated soil environment; ④ deploying soil contamination governance and remediation; ⑤ elevating the capacity of soil environment monitoring and management; ⑥ speeding up the construction of soil environment protection engineering. Generally, the targets of the work arrangements can be summarized as to understand the status quo of soil environment throughout the nation by 2015, and to establish soil environment protection system so as to significantly improve the soil environment quality by 2020.

In addition, the development of Chinese modern agriculture has gradually undergone a significant transformation to more environmentally friendly utilizations, such as organic agriculture, ecological agriculture, leisure agriculture, and agricultural tourism, in addition to modernized intensive agriculture, where the applications of chemical fertilizers and pesticides have been greatly reduced through scientific and effective management. In addition, the *Action Plan for Zero Growth of Chemical Fertilizers Consumption by 2020* and *the Action Plan for Zero Growth of Pesticides Consumption by 2020* were promulgated by the former MOA on February 17, 2015, so as to promote the decrement of chemical fertilizers and pesticides consumption and to dynamically explore the strategic approach to modernized sustainable agriculture with high efficiency, high quality, safety products, energy saving, and resources conservation, environment. Within the *Action Plan for Zero Growth of Chemical Fertilizers Consumption by 2020*, various key tasks were identified, including: ① promoting soil testing and formula fertilization technology for diverse crops; ② modifying the fertilization patterns; ③ application of new fertilizers and new technologies; ④ encouraging the utilization of organic fertilizers resources; ⑤ enhancing the quality of cultivated land. Within the *Action Plan for Zero Growth of Pesticides Consumption by 2020*, several major tasks were deployed, such as ① establishing monitoring of diseases and pests as well as a warning system; ② encouraging scientific and rational pesticide applications; ③ promoting green technologies for the prevention and control of pests; ④ facilitating integrated prevention and control of pests.

Environmental taxation (or ecological taxation, green taxation) is an effective economic means to reflect the social costs of environmental pollution and ecological destruction to production costs and market prices, so as to allocate environmental resources through market mechanism. Environmental taxation has been applied to many developed countries to acquired precious experience and lessons. However, due to diverse national conditions and taxation policies, the environmental taxation systems in various countries can be quite different. In many developing countries, environmental taxation has become an important policy to advocate environmental protection and to correct market failure. In China, the initiative of environmental taxation was proposed by the MOF in December 2011. In June 2015, *the Exposure Draft of the Environmental Protection Taxation* was publicized by the State Council. And on December 25, 2016, *the Law of the People's Republic of China on Environmental Protection Taxation* was finally promulgated and became effective on January 1, 2018, which is the special law to embody green taxation. Environmental protection taxation, basically evolved from the pollution emission charges, will enhance the ecological environment protection through the regulation and control of taxation, to form the effective binding and incentive mechanism, and to fulfill the environmental liability of polluters.

References

Liu, Y., Xia, X., & Song, J. (1999). Chinese land resources and its sustainable utilizations. *Management on Geological Science and Technology, 1*, 1–6.

Ma, H. (2005). *Research on protection and sustainable utilization for land resource in China*. Master: Soochow University.

MEE. (2018). *The 2017 bulletin of Chinese ecology and environment status quo*. Beijing: MEE, Ministry of Ecology and Environment.

MEP. (2017). *The 2016 bulletin of Chinese environment status quo*. Beijing: MEP, Ministry of Environmental Protection.

MEP and MLR. (2014). *The bulletin of the first investigation on nationwide soil pollution status quo*. Beijing: MEP and MLR.

MNR. (2018). *The 2017 statistical bulletin of Chinese land*. Beijing: Minerals and Marine Resources. MNR, Ministry of Natural Resources.

MWR and NBS (2013). *The bulletin of the first national census for water* (p. 22). Beijing: MWR and NBS, China Water & Power Press.

SFA (2015). *The bulletin of the fifth national desertification and sandification monitoring* (p. 13). Beijing: SFA, State Forestry Administration.

Song, W., Chen, B., & Liu, L. (2013). Soil heavy metal pollution of cultivated land in China. *Research of Soil and Water Conservation, 20*(2), 293–298.

Sun, X. (2015). To envisage status quo of land resources in China and to perfect land management system. *Jilin Agriculture, 23*, 119.

Zang, C., & Li, Y. (2018). Study on the status quo of soil pollution, control and remediation in China. *Land & Resources, 4*, 48–49.

Zhuang, G. (2015). Current situation of national soil pollution and strategies on prevention and control. *Bulletin of Chinese Academy of Sciences, 30*(4), 477–483.

Comparison Table

1.1 Institutions & Organizations

the Leading Group of Environmental Protection	环境保护领导小组 (1974—1982)
the National People's Congress	全国人民代表大会 (1954—)
the Ministry of Urban and Rural Construction and Environmental Protection	城乡建设环境保护部 (1982—1988)
the Environmental Protection Commission of the State Council	国务院环境保护委员会 (1984—)
the National Environmental Protection Administration	国家环境保护局 (1984—1998)
the State Environmental Protection Administration	国家环境保护总局 (1998—2008)
the Ministry of Environmental Protection	环境保护部 (2008—2018)
the Ministry of Ecology and Environment	生态环境部 (2018—)
the Communist Party of China	中国共产党 (1921—)
the Central Committee of the Communist Party of China	中国共产党中央委员会 (1927—)
the National Congress of the Communist Party of China	中国共产党全国代表大会 (1921—)
the Third Plenary Session of the Sixteenth Central Committee of the CPC	中国共产党第十六届中央委员会第三次全体会议 (2003)
the Organization Department of the CCCPC	中国共产党中央委员会组织部 (1924—)
the Chinese People's Political Consultative Conference	中国人民政治协商会议 (1949—)

the National Development and Reform Commission	国家发展和改革委员会 (2003—)
the Ministry of Supervision	监察部 (1954—2018)
the National Bureau of Statistics	国家统计局 (1952—)
the Ministry of Water Resources	水利部 (1949—)
the Ministry of Housing and Urban-Rural Development	住房和城乡建设部 (2008—)
the Ministry of Agriculture	农业部 (1949—2018)
the Ministry of Agriculture and Rural Affairs	农村农业部 (2018—)
the Ministry of Land and Resources	国土资源部 (2008—2018)
the Ministry of Natural Resources	自然资源部 (2018—)
the State Price Bureau	国家物价局 (1982—1994)
the State Forestry Administration	国家林业局 (2008—2018)
the State Forestry and Grassland Administration	国家林业和草原局 (2018—)
the Ministry of Transport	交通运输部 (2008—)
the Ministry of Industry and Information Technology	工业和信息化部 (2008—)
the Ministry of Finance	财政部 (1949—)
the Ministry of Health	卫生部 (1998—2013)
the National Health and Family Planning Commission	国家卫生和计划生育委员会 (2013—2018)
the National Medical Security Bureau	国家医疗保障局 (2018—)
the State Flood Control and Drought Relief Headquarter	国家防汛抗旱总指挥部 (1992—)
the State Planning Commission	国家计划委员会 (1952—1998)
the State Construction Commission	国家建设委员会 (2018—)
the State Economy Commission	国家经济委员会 (1956—1988)
the State Development Planning Commission	国家发展计划委员会 (1998—2003)
the State Economy and Trade Commission	国家经济贸易委员会 (1993—2003)
the Government Administration Council of the Central People's Government	中央人民政府政务院 (1949—1954)

1.2 National Conference on Environmental Protection

the First National Conference on Environmental Protection	第一次全国环境保护会议 (1973 年 8 月)
the Second National Conference on Environmental Protection	第二次全国环境保护会议 (1983 年 12 月)
the Third National Conference on Environmental Protection	第三次全国环境保护会议 (1989 年 4 月)
the Fourth National Conference on Environmental Protection	第四次全国环境保护会议 (1996 年 7 月)
the Fifth National Conference on Environmental Protection	第五次全国环境保护会议 (2002 年 1 月)
the Sixth National Conference on Environmental Protection	第六次全国环境保护会议 (2006 年 4 月)
the Seventh National Conference on Environmental Protection	第七次全国环境保护会议 (2011 年 12 月)
the Eighth National Conference on Ecological and Environmental Protection	第八次全国生态环境保护会议 (2018 年 5 月)

1.3 The Five-Year Plans

the First Five-Year Plan of National Economic and Social Development	一五计划 (1953—1957)
the Second Five-Year Plan of National Economic and Social Development	二五计划 (1958—1962)
the Third Five-Year Plan of National Economic and Social Development	三五计划 (1966—1970)
the Fourth Five-Year Plan of National Economic and Social Development	四五计划 (1971—1975)
the Fifth Five-Year Plan of National Economic and Social Development	五五计划 (1976—1980)
the Sixth Five-Year Plan of National Economic and Social Development	六五计划 (1981—1985)
the Seventh Five-Year Plan of National Economic and Social Development	七五计划 (1986—1990)

the Eighth Five-Year Plan of National Economic and Social Development	八五计划 (1991—1995)
the Ninth Five-Year Plan of National Economic and Social Development	九五计划 (1996—2000)
the Tenth Five-Year Plan of National Economic and Social Development	十五计划 (2001—2005)
the Eleventh Five-Year Plan of National Economic and Social Development	十一五计划 (2006—2010)
the Twelfth Five-Year Plan of National Economic and Social Development	十二五计划 (2011—2015)
the Thirteenth Five-Year Plan of National Economic and Social Development	十三五计划 (2016—2020)
the Outlines of the Ninth Five-Year Plan for the National Economic and Social Development of the People's Republic of China and the Long-Term Objectives for 2010	《国民经济和社会发展"九五"计划和2010年远景目标纲要》(1996)
the Outlines of the Tenth Five-Year Plan for the National Economic and Social Development of the People's Republic of China	《中华人民共和国国民经济和社会发展第十个五年计划纲要》(2001)
the Outlines of the Eleventh Five-Year Plan for the National Economic and Social Development of the People's Republic of China	《中华人民共和国国民经济和社会发展第十一个五年计划纲要》(2006)
the Outlines of the Thirteenth Five-Year Plan for the National Economic and Social Development of the People's Republic of China	《中华人民共和国国民经济和社会发展第十三个五年计划纲要》(2016)

1.4　Laws

the Constitution of the People's Republic of China	《中华人民共和国宪法》(1954)
the Law of the People's Republic of China on Environmental Protection (on Trial)	《中华人民共和国环境保护法(试行)》(1979)
the Law of the People's Republic of China on Environmental Protection	《中华人民共和国环境保护法》(1989)
the Law of the People's Republic of China on Water Pollution Prevention and Control	《中华人民共和国水污染防治法》(1984)
the Law of the People's Republic of China on Air Pollution Prevention and Control	《中华人民共和国大气污染防治法》(1987)

the Law of the People's Republic of China on Marine Environment Protection	《中华人民共和国海洋环境保护法》(1982)
the Law of the People's Republic of China on Solid Waste Pollution Prevention and Control	《中华人民共和国固体废物污染环境防治法》(1995)
the Law of the People's Republic of China on Environmental Noise Pollution Prevention and Control	《中华人民共和国环境噪声污染防治法》(1996)
the Law of the People's Republic of China on Land Desertification Prevention and Control	《中华人民共和国防沙治沙法》(2001)
the Law of the People's Republic of China on Promoting Cleaner Production	《中华人民共和国清洁生产促进法》(2002)
the Law of the People's Republic of China on Environmental Impact Assessment	《中华人民共和国环境影响评价法》(2002)
the Law of the People's Republic of China on Radioactive Pollution Prevention and Control	《中华人民共和国放射性污染防治法》(2003)
the Law of the People's Republic of China on Renewable Energy	《中华人民共和国可再生能源法》(2005)
the Law of the People's Republic of China on Promoting Circular Economy	《中华人民共和国循环经济促进法》(2008)
the General Provisions of the Civil Law of the People's Republic of China	《中华人民共和国民法通则》(1986)
the Civil Procedural Law of the People's Republic of China	《中华人民共和国民事诉讼法》(1991)
the Law of the People's Republic of China on Liability for Tort	《中华人民共和国侵权责任法》(2009)
the Criminal Law of the People's Republic of China	《中华人民共和国刑法》(1997)
the Law of the People's Republic of China on Energy Conservation	《中华人民共和国节约能源法》(1997)
the Law of the People's Republic of China on Water	《中华人民共和国水法》(1988)
the Law of the People's Republic of China on Soil Pollution Prevention and Control	《中华人民共和国土壤污染防治法》(2018)
the Law of the People's Republic of China on Environmental Protection Tax	《环境保护税法》(2016)
the Law of the People's Republic of China on Tax Collection and Management	《中华人民共和国税收征收管理法》(1992)
the Law of the People's Republic of China on Legislation	《中华人民共和国立法法》(2000)

1.5 Environmental Policies and Plans

the Opinions on Accelerating the Advancement of the Ecological Civilization Construction	中共中央国务院《关于加快推进生态文明建设的意见》(2015)
China's Agenda 21 -- White Paper on China's Population, Environment and Development in 21st Century	《中国 21 世纪议程——中国 21 世纪人口、环境与发展白皮书》(1994)
the Outlines of National Environmental Protection Work from 1993 to 1998	《全国环境保护工作纲要 (1993—1998)》
the Ten Strategies on Environment and Development	《中国环境与发展十大对策》(1992)
the General Planning for the Institutional Reform of Ecological Civilization	《生态文明体制改革总体方案》(2015)
the Index System of Green Development	《绿色发展指标体系》(2016)
the Objectives System of Ecological Civilization Construction Auditing	《生态文明建设考核目标体系》(2016)
the Ecological and Environmental Protection Plan during the 13th FYP	《"十三五"生态环境保护规划》(2016)
Some Preliminary Opinions Concerning Strengthening the Guidance on Environmental Protection Plan	《关于加强环境保护计划指导的几点初步意见》(1982)
the Decisions on Several Issues Concerning Environmental Protection	《关于环境保护若干问题的决定》(1996)
Some Regulations Concerning Environmental Protection and Improvement (on Trial Draft)	《关于保护和改善环境的若干规定（试行草案）》(1973)
the Report on Strengthening the Environmental Protection Work	《关于加强环境保护工作的报告》(1976)
the Regulations on the Design of Environmental Protection of Construction Project	《建设项目环境保护设计规定》(1987)
the Decision on Implementing the Scientific Outlook on Development and Strengthening Environmental Protection	国务院《关于落实科学发展观加强环境保护的决定》(2005)
Opinions Concerning Strengthening the Key Works of Environmental Protection	国务院《关于加强环境保护重点工作的意见》(2011)
the Decision on Several Major Issues Concerning the Comprehensively Deepening Reform	中共中央《关于全面深化改革若干重大问题的决定》(2013)
the Decisions on further Enhancing the Environmental Protection Work	国务院《关于进一步加强环境保护工作的决定》(1990)
the General Planning for Ecological Civilization System Reform	《生态文明体制改革总体方案》(2015)

the Key Special Plans for Ecological Construction and Environmental Protection in the 10th FYP	《"十五"生态建设和环境保护重点专项规划》(2001)
the Key Engineering Projects of National Environmental Protection Plan during the 10th FYP	《国家环境保护"十五"重点工程项目规划》(2002)
the National Ecological and Environmental Protection Plan during the 10th FYP	《全国生态环境保护"十五"计划》(2002)
the Science and Technology Development Plan for National Environmental Protection during the 10th FYP	《全国生态环境保护科技发展"十五"计划》(2002)
the National Environmental Monitoring Plan during the 10th FYP	《全国环境监测"十五"计划纲要》(2002)
the National Environmental Protection Plan during the 12th FYP	《国家环境保护"十二五"规划》(2011)
National Major Function Zoning Plan	《全国主体功能区规划》(2010)
the Compiling Technical Outlines of the Local Environmental Protection Plan during the Tenth Five-Year Plan and the Long-Term Objectives for 2015	《地方环境保护"十五"计划和2015年长远目标纲要编制技术大纲》(2000)
the Plan for the Disintegration of Total Emission Cap on Major Pollutants during the 10th FYP	《"十五"期间全国主要污染物排放总量控制分解计划》(2001)
the National Environmental Protection Plan during the 10th FYP	《国家环境保护"十五"规划》(2001)
the Environmental Protection Plan for the Pearl River Delta Region	《珠江三角洲区域环境保护规划》(2002)
the Environmental Protection Plan for Guangdong Province	《广东省环境保护规划》(2005)
the Environmental Protection Plan for the Yangtze River Delta Region	《长江三角洲区域环境保护规划》(2005)
the Key Points of Environmental Protection Work Report	《环境保护工作汇报要点》(1978)
the Decisions on Strengthening the Environmental Protection Work During the Period of National Economy Adjustment	《关于在国民经济调整时期加强环境保护工作的决定》(1981)
the Index System of Green Development	《绿色发展指标体系》(2016)
the Objectives System of Ecological Civilization Construction Auditing	《生态文明建设考核目标体系》(2016)
the Compiling Technical Outlines of the Plan for Comprehensive Urban Environment Governance	《城市环境综合整治规划编制技术大纲》(1993)
the Guidelines for Environmental Planning	《环境规划指南》(1994)
the Ecological and Environmental Protection Plan during the 13th FYP	《"十三五"生态环境保护规划》(2016)

1.6 Emission Charges & Environmental Tax

The Interim Measures of Emissions Charges Collection	《征收排污费暂行办法》(1982)
The Measures of the Financial Management and Accounting for the Emissions Charges	《征收超标准排污费财务管理和会计核算办法》(1984)
The Interim Measures of the Compensable Utilizations of the Special Funds for Pollution Source Governance	《污染源治理专项基金有偿使用暂行办法》(1988)
the Notice on the Implementation of the Pilot Project to Collect Emissions Charges of Sulfur Dioxide from Industrial Coal Burning	《关于开展征收工业燃煤二氧化硫排污费试点工作的通知》(1992)
the Rescriptum on the Issues Concerning the Expansion of the Pilot Project to Collect Emissions Charges of Sulfur Dioxide	《关于二氧化硫排污收费扩大试点工作有关问题的批复》(1996)
the Notice on the Expansion of the Pilot Project to Collect Emissions Charges of Sulfur Dioxide in Acid Rain Control Zones and Sulfur Dioxide Pollution Control Zones	《关于在酸雨控制区和二氧化硫污染控制区开展征收二氧化硫排污费扩大试点的通知》(1998)
the Notice on the Implementation of Charge Standards in the Pilot Project to Collect Emissions Charges of Sulfur Dioxide	《关于二氧化硫排污收费试点征收标准执行问题的通知》(2000)
the Notice on Collecting the Emissions Charges of Sewage	《关于征收污水排污费的通知》(1993)
the Zoning Schemes of the Acid Rain Control Zone and Sulfur Dioxide Pollution Control Zone	《酸雨控制区和二氧化硫污染控制区划分方案》(1998)
the Notice on the Demonstration Work Arrangement for Total Emission Cap for Sulfur Dioxide and Emission Trading Policy	《关于二氧化硫排放总量控制及排污权交易政策实施示范工作安排的通知》(2002)
the Ordinances of Management for the Collection and Utilizations of Emission Charges	《排污费征收使用管理条例》(2003)
the Measures of Management for the Emissions Charges Standards	《排污费征收标准管理办法》(2003)
the Notice on the Issues of Adjusting the Emissions Charges Standards	《关于调整排污费征收标准等有关问题的通知》(2014)
the Special Plan for the Science and Technology of Blue Sky Project during the 12th FYP	《蓝天科技工程"十二五"专项规划》(2012)
the Notification Concerning further Perfecting the Air Quality Monitoring and Early Warning Work under the Weather Conditions for Potential Heavy Air Pollution	《关于进一步做好重污染天气条件下空气质量监测预警工作的通知》(2013)

the Comprehensive Emission Standards of Air Pollutants	《大气污染物综合排放标准》(1996)
the Specifications of Ambient Air Quality Monitoring (on Trial)	《环境空气质量监测规范》(试行)(2007)
the Interim Ordinances of the Resources Taxes	《资源税暂行条例》(1993)
the Amendment of the Interim Ordinances of the Resources Taxes	《资源税暂行条例》修订案 (2011)
the Notice Concerning Promoting the Overall Reform on Resources Taxes	《关于全面推进资源税改革的通知》(2016)
the Exposure Draft of the Environmental Protection Taxation	《环境保护税法(征求意见稿)》(2015)
the Regulations of Management for the Pollutant Emission Reporting and Registration	《排放污染物申报登记管理规定》(1992)
the Notice on Comprehensive Implementation of Emission Reporting and Registration	《关于全面推行排污申报登记的通知》(1997)
the Implementation Measures of Controlling the Emission Permit System to Control Pollutant Emission	《控制污染物排放许可制实施方案》(2016)
the Guiding Opinions Concerning further Promoting the Pilot Program of the Compensable Utilizations of the Emission Permit and Emissions Trading	《关于进一步推进排污权有偿使用和交易试点工作的指导意见》(2014)
the Interim Measures of Management for the Income from Transferring the Emission Permit	《排污权出让收入管理暂行办法》(2015)

1.7 EIA, Public Participation & Information Disclosure

the Notice Concerning Well Preparation for the Preliminary Work of Infrastructure Construction	《关于做好基本建设前期工作的通知》(1979)
the Measures of Management for Environmental Protection of Infrastructure Construction Project	《基本建设项目环境保护管理办法》(1981)
the Measures of Management for Environmental Protection of Construction Project	《建设项目环境保护管理办法》(1986)
the Measures of Management for the Certificate System of Environmental Impact Assessment of Construction Project (on Trial)	《建设项目环境影响评价证书管理办法(试行)》(1986)
the Measures of Management for the Qualification and Certificate System of Environmental Impact Assessment of Construction Project	《建设项目环境影响评价资质证书管理办法》(1999)

the Ordinances of Management for Environmental Protection of Construction Project	《建设项目环境保护管理条例》(1998)
the Catalogue of Classification Management for Environmental Protection of Construction Project (on Trial)	《建设项目环境保护分类管理名录（试行）》(1999)
the Catalogue of Classification Management for Environmental Protection of Construction Project	《建设项目环境保护分类管理名录》(2002)
the Catalogue of Classification Management for Environmental Impact Assessment of Construction Project	《建设项目环境影响评价分类管理名录》(2008)
the Measures of Management for the Completion and Acceptance of Environmental Protection of Construction Projects	《建设项目竣工环境保护验收管理办法》(2001)
the Interim Measures of the Completion and Acceptance of Environmental Protection of Construction Projects	《建设项目竣工环境保护验收暂行办法》(2017)
the Technical Guidelines for Environmental Impact Assessment for Plans (on Trial)	《规划环境影响评价技术导则（试行）》(2003)
the Measures of Management for the Reviewing Experts Database of Environmental Impact Assessment	《环境影响评价审查专家库管理办法》(2003)
the Measures of Reviewing Environmental Impact Assessment Report of Special Plan	《专项规划环境影响报告书审查办法》(2003)
Scope of the Plans to Prepare the Environmental Impact Report (on Trial)	《编制环境影响报告书的规划的具体范围（试行）》(2004)
Scope of the Plans to Prepare the Environmental Impact Chapters (on Trial)	《编制环境影响篇章或说明的规划的具体范围（试行）》(2004)
the Measures of Management for the Qualification and Certificate System of Environmental Impact Assessment of Construction Project	《建设项目环境影响评价资质管理办法》(2005)
the Interim Measures of Public Participation in Environmental Impact Assessment	《环境影响评价公众参与暂行办法》(2006)
the Measures of Public Participation in Environmental Impact Assessment	《环境影响评价公众参与办法》(2018)
the Implementation Procedures of Environmental Impact Assessment Reform during the 13th FYP	《"十三五"环境影响评价改革实施方案》(2016)
the Notice on Strengthening the Management for EIA for Construction Projects Loaned by the International Financial Organizations	《关于加强国际金融组织贷款建设项目环境影响评价管理工作的通知》(1993)

the Ordinances of the People's Republic of China on Government Information Disclosure	《中华人民共和国政府信息公开条例》(2007)
the Measures of Environmental Information Disclosure (on Trial)	《环境信息公开办法（试行）》(2007)
the Guidelines for Disclosure of Government Information on EIA for Construction Project (on Trial)	《建设项目环境影响评价政府信息公开指南（试行）》(2013)
the Interim Regulations on Environmental Management for Economic Zones Open to the Outside World	《对外经济开放地区环境管理暂行规定》(1986)
Several Opinions Concerning to further Improve the Environmental Protection and Management for Construction Project	《关于进一步做好建设项目环境保护管理工作的几点意见》(1993)
the Measures of Reviewing the Environmental Impact Assessment Report of Special Plan	《专项规划环境影响报告书审查办法》(2003)
the Interim Regulations on the System of EIA Engineers Vocational Certificate	《环境影响评价工程师职业资格制度暂行规定》(2004)
the Implementation Measures of the Examination of EIA Engineer Vocational Certificate	《环境影响评价工程师职业资格考试实施办法》(2004)
the Measures of the Review of EIA Engineer Vocational Certificate	《环境影响评价工程师职业资格考核认定办法》(2004)
the Measures of Public Participation in Environmental Protection	《环境保护公众参与办法》(2015)
the Guiding Opinions Concerning Cultivating and Regulating the Orderly Development of Environmental Non-Governmental Organizations	《关于培育引导环保社会组织有序发展的指导意见》(2010)
the Guiding Opinions Concerning Promoting Public Participation in Environmental Protection	《关于推进环境保护公众参与的指导意见》(2014)
the Notice on Strengthening the Information Disclosure of Environmental Supervision on Pollution Sources	《关于加强污染源环境监管信息公开工作的通知》(2013)
the Catalogue of Information Disclosure of Environmental Supervision on Pollution Sources (the First Batch)	《污染源环境监管信息公开目录（第一批）》(2013)
the Interim Measures of Enterprises and Public Institutions to Disclose Environmental Information	《企业事业单位环境信息公开暂行办法》(2014)
the Guiding Opinions Concerning Promoting Public Participation in Environmental Protection	《关于推进环境保护公众参与的指导意见》(2014)
the Interim Ordinances of Defending State Secrets	《保守国家机密暂行条例》(1951)

1.8 Emission Cap

the Measures of Performance Evaluation on Total Emission Reduction of Major Pollutants during the 12th FYP	《"十二五"主要污染物总量减排考核办法》(2013)
the Emission Standards of Industrial "Three-Waste" (on Trial)	《工业三废排放试行标准》(GBJ 4—1973)
the Notice on the Comprehensive Schemes of Work for Energy Conservation and Emission Reduction	《关于印发节能减排综合性工作方案的通知》(2007)
the Notification on the Construction and Operation of Wastewater Treatment Facilities in Cities and Towns in the First Half of 2017	《关于2017年上半年全国城镇污水处理设施建设和运行情况的通报》(2017)
the Bulletin of Urban and Rural Construction Statistics of 2014	《2014年城乡建设统计公报》(2015)
the Opinions Concerning Deeply Promoting the Construction of New Type of Urbanization	《国务院关于深入推进新型城镇化建设的若干意见》(2016)
the Plan for Emissions Cap of Major Pollutants Nationwide During the 9th FYP	《"九五"期间全国主要污染物排放总量控制计划》(1996)
the Ninth Five-Year Plan for the National Economic and Social Development of the People's Republic of China and the Long-Term Objectives for 2010	《国民经济和社会发展"九五"计划和2010年远景目标》(1996)
the Notice on Several Issues Concerning the Allowance Allocation of the Emissions Cap of Major Pollutants from Construction Projects	《关于核定建设项目主要污染物排放总量控制指标有关问题的通知》(2003)
the Notice on the Scheme of Determining the Environmental Capacity of Surface Water and Ambient Air Nationwide	《关于印发全国地表水环境容量和大气环境容量核定工作方案的通知》(2003)
the Measures of Monitoring the Emissions Reduction of Major Pollutants	《主要污染物总量减排监测办法》(2007)
the Measures of Evaluating the Emissions Reduction of Major Pollutants during the 11th FYP (on Trial)	《"十一五"主要污染物总量减排核查办法（试行）》(2007)
the Detailed Rules of Auditing the Emissions Reduction of Major Pollutants	《主要污染物总量减排核算细则（试行）》(2007)
the Notice Concerning the Implementation of Quarterly Reporting System for Emissions Reduction of Major Pollutants	《关于实施"十一五"主要污染物总量减排措施季度报告制度的通知》(2007)

1.9 Redline

the National Ecological Redline - the Technical Guidelines for the Demarcation of Ecological Function Zone Redline (on Trial)	《国家生态保护红线 —— 生态功能基线划定技术指南（试行）》(2014)
the Technical Guidelines for the Demarcation of Ecological Redline	《生态保护红线划定技术指南》(2015)
the Opinions Concerning Demarcating and Safeguarding the Ecological Redline	《关于划定并严守生态保护红线的若干意见》(2017)
the Outlines of the General Planning for National Land Utilizations (2006—2020)	《全国土地利用总体规划纲要（2006—2020年）》(2008)
the Decisions on Accelerating the Reform and Development of Water Conservancy	《关于加快水利改革发展的决定》(2010)
the Opinions on Implementing the Strictest Water Resource Management System	《关于实行最严格水资源管理制度的意见》(2012)
the Law of the People's Republic of China on Tort Liability	《中华人民共和国侵权责任法》(2009)
the Law of the People's Republic of China on Property Rights	《中华人民共和国物权法》(2007)
the Pilot Scheme of the Reform of the Environmental Damage Compensation System	《生态环境损害赔偿制度改革试点方案》(2015)
the Recommended Methods of Calculating the Amount of Damage Caused by Environmental Pollution (Version I)	《环境污染损害数额计算推荐方法（第Ⅰ版）》(2011)
the Recommended Methods of the Validation and Assessment of Environmental Damage (Version II)	《环境损害鉴定评估推荐方法（第Ⅱ版）》(2014)
the Notice of Emergency Response to the National Emergent Environmental Incidents	《国家突发环境事件应急预案》(2014)
the Recommended Methods of Environmental Damage Assessment during the Emergency Response to the Unforeseen Environmental Incidents	《突发环境事件应急处置阶段环境损害评估推荐方法》(2014)

1.10 Cleaner Production

the Notice Concerning Demonstration Pilot Site Project for Cleaner Production Implementation	《关于实施清洁生产示范试点计划的通知》(1999)
the Some Regulations Concerning Technical Modification for Industrial Pollution Prevention	国务院《关于结合技术改造防治工业污染的几项规定》(1983)
the Some Opinions Concerning Promoting Cleaner Production	《关于促进清洁生产的一些意见》(1997)
the Catalogue of Technical Guidance of Cleaner Production for Key Enterprises (the 1st Batch)	《国家重点行业清洁生产技术导向目录（第一批）》(2000)
the Catalogue of Technical Guidance of Cleaner Production for Key Enterprises (the 2nd Batch)	《国家重点行业清洁生产技术导向目录（第二批）》(2003)
the Catalogue of Technical Guidance of Cleaner Production for Key Enterprises (the 3rd Batch)	《国家重点行业清洁生产技术导向目录（第三批）》(2006)
the Catalogue of Outdated Production Capacity, Technologies and Products to be Phased-Out (the 1st Batch)	《淘汰落后生产能力、工艺和产品的目录（第一批）》(1999)
the Catalogue of Outdated Production Capacity, Technologies and Products to be Phased-Out (the 2nd Batch)	《淘汰落后生产能力、工艺和产品的目录（第二批）》(1999)
the Catalogue of Outdated Production Capacity, Technologies and Products to be Phased-Out (the 3rd Batch)	《淘汰落后生产能力、工艺和产品的目录（第三批）》(2002)
the Some Opinions Concerning Fully Implementing the Law of People's Republic of China on Promoting Cleaner Production	《关于贯彻落实〈清洁生产促进法〉的若干意见》(2003)
the Some Opinions Concerning Expediting The Implementation of Cleaner Production	《关于加快推行清洁生产的意见》(2003)
the Interim Measures of Cleaner Production Audit	《清洁生产审核暂行办法》(2004)
the Measures of Cleaner Production Audit	《清洁生产审核办法》(2016)
the Measures of Utilizations and Management for the Cleaner Production Special Funds Subsidized from Central to Local Governments	《中央补助地方清洁生产专项资金使用管理办法》(2004)
the Regulations on the Procedures of Cleaner Production Audit for Key Enterprises	《重点企业清洁生产审核程序的规定》(2005)

the Catalogue of Toxic and Hazardous Materials Requiring Cleaner Production Audit (the 1st Batch)	《需重点审核的有毒有害物质名录（第一批）》(2005)
the Notice Concerning Further Enhancement on Cleaner Production Audit for Key Enterprises	《关于进一步加强重点企业清洁生产审核工作的通知》(2008)
the Catalogue of Toxic and Hazardous Materials Requiring Cleaner Production Audit (the 2nd Batch)	《需重点审核的有毒有害物质名录（第二批）》(2008)
the Implementation Guidance on Assessment and Acceptance of Cleaner Production Audit for Key Enterprises (on Trial)	《重点企业清洁生产审核评估、验收实施指南（试行）》(2008)
the Notice Concerning Enhancement on Promoting Cleaner Production for Industries and Information Technology Sectors	《关于加强工业和通信业清洁生产促进工作的通知》(2009)
the Interim Measures of Management for the Central Finance Cleaner Production Special Funds	《中央财政清洁生产专项基金管理暂行办法》(2009)
the Notice Concerning Further Promoting Cleaner Production for Key Enterprises	《关于深入推进重点企业清洁生产的通知》(2010)
the Catalogue of Business Classification Management for Cleaner Production for Key Enterprises	《重点企业清洁生产行业分类管理名录》(2010)
the General Provisions of Compiling the Cleaner Production Assessment Indicator System (on Trial)	《清洁生产评价指标体系编制通则（试行）》(2013)
the Implementation Procedures of Cleaner Production Technologies for Key Enterprises and Industries in Air Pollution Prevention and Control	《大气污染防治重点工业行业清洁生产技术推行方案》(2014)
the Standards of Cleaner Production Audit for Industries	《工业清洁生产审核规范》(2015)
the Standards of Effectiveness Evaluation of Cleaner Production Audit for Industries	《工业清洁生产实施效果评估规范》(2015)
the Announcement of Key Enterprises to Pass Cleaner Production Audit (the 1st Batch)	《实施清洁生产审核并通过评估验收的重点企业名单（第一批）》(2010)
the Announcement of Key Enterprises to Pass Cleaner Production Audit (the 2nd Batch)	《实施清洁生产审核并通过评估验收的重点企业名单（第二批）》(2010)
the Announcement of Key Enterprises to Pass Cleaner Production Audit (the 3rd Batch)	《实施清洁生产审核并通过评估验收的重点企业名单（第三批）》(2011)
the Announcement of Key Enterprises to Pass Cleaner Production Audit (the 4th Batch)	《实施清洁生产审核并通过评估验收的重点企业名单（第四批）》(2011)
the Announcement of Key Enterprises to Pass Cleaner Production Audit (the 5th Batch)	《实施清洁生产审核并通过评估验收的重点企业名单（第五批）》(2012)

1.11 Ecological Compensation

the Ordinances on Returning Farmland to Forest	《退耕还林条例》(2002)
the Interim Measures of Forest Ecological Benefit Subsidy Management	《森林生态效益补助资金管理办法（暂行）》(2001)
the Measures of the Central Forest Ecological Benefits Compensation Fund Management	《中央森林生态效益补偿基金管理办法》(2004)
the Measures of the Central Finance Forest Ecological Benefits Compensation Fund Management	《中央财政森林生态效益补偿基金管理办法》(2007)
Some Opinions Concerning Strengthening Grassland Protection and Construction	《关于加强草原保护与建设的若干意见》(2002)
the Law of the People's Republic of China on Grassland	《中华人民共和国草原法》(1985)
the National Master Plan for Grassland Conservation, Construction and Utilizations	《全国草原保护建设利用总体规划》(2007)
the Master Plan for the Ecological Protection and Construction of National Nature Reserves at Three Rivers Sources District, Qinghai Province	《青海三江源国家自然保护区生态保护和建设总体规划》(2005)
the Law of the People's Republic of China on Mineral Resources	《中华人民共和国矿产资源法》(1986)
the Law of the People's Republic of China on Fisheries	《中华人民共和国渔业法》(1986)
the Ordinances on Wetland Protection Poyang Lake	《鄱阳湖湿地保护条例》(2003)
the Measures of Pursuing the Liability of Party and Government Leading Cadres for Ecological and Environmental Damage (on Trial)	《党政领导干部生态环境损害责任追究办法（试行）》(2015)

1.12 Air Pollution

the Implementation Provisions for the Law of the People's Republic of China on Air Pollution Prevention and Control	《大气污染防治法实施细则》(1991)
the Ambient Air Quality Standards	《环境空气质量标准》(1982)
the Interim Measures of Developing Coal Briquette for Civilian Applications	《关于发展民用型煤的暂行办法》(1987)

the Measures of Management for Smoke and Dust Control in Cities	《城市烟尘控制区管理办法》(1987)
the Measures of Management for Supervision on Motor Vehicle Exhaust Pollution	《汽车排气污染监督管理办法》(1990)
the Norm of Ambient Air Quality Monitoring (on Trial)	《环境空气质量监测规范》(试行)(2007)
the Amendment of Ambient Air Quality Standards (2012)	《环境空气质量标准》(修订)(2012)
the Action Plan for Air Pollution Prevention and Control	《大气污染防治行动计划》(2013)
the Compilation of Advanced Technologies of Air Pollution Prevention and Control	《大气污染防治先进技术汇编》(2014)
the Policy on Pollution Prevention and Control Technologies of SO_2 Emission from Coal Burning	《燃煤 SO_2 排放污染防治技术政策》(2002)
the Amendment of the Air Pollutants Emission Standards for Thermal Power Plant	《火电厂大气污染物排放标准》修订案 (2003)
the Technical Guidelines for Air Pollution Governance Engineering	《大气污染治理工程技术导则》(2010)
the Amendment of the Air Pollutants Emission Standards for Thermal Power Plant	《火电厂大气污染物排放标准》修订案 (2011)
the Policy on Integrated Pollution Prevention and Control Technologies for Fine Particulate Matter	《环境空气细颗粒物污染综合防治技术政策》(2013)
the Pollution Prevention and Control Plan of the Two-Control-Zone "Acid Rain Control Zone and Sulfur Dioxide Control Zone" during the 10th FYP	《两控区酸雨和二氧化硫污染防治"十五"计划》(2002)
the Schemes of Designating Key Cities for Air Pollution Prevention and Control	《大气污染防治重点城市划定方案》(2002)
the Measures of Management for Electricity Price of Coal Fired Power Generator with Desulfurization Facility and the Operation of Desulfurization Facility (on Trial)	《燃煤发电机组脱硫电价及脱硫设施运行管理办法》(试行)(2007)
the Guiding Opinions Concerning the Promotion of Joint Regional Air Pollution Prevention and Control Work for Improving Regional Air Quality	《关于推进大气污染联防联控工作改善区域空气质量的指导意见》(2010)
the Air Pollution Prevention and Control for Key Regions during the 12th FYP	《重点区域大气污染防治"十二五"规划》(2012)
the Action Plan for Energy Conservation, Emission Reduction and Low Carbon Development between 2014 and 2015	《2014—2015 年节能减排低碳发展行动方案》(2014)

  Three-Year (2018—2020) Action Plan for Winning the Blue-Sky War   《打赢蓝天保卫战三年行动计划》(2018)

  the Measures of Performance Evaluation on the Implementation of the Action Plan for Air Pollution Prevention and Control (on Trial)   《大气污染防治行动计划实施情况考核办法（试行）》(2014)

  the Regulations on the Implementation of the Law of the People's Republic of China on Air Pollution Prevention and Control   《中华人民共和国大气污染防治法实施细则》(1991)

  the Technical Guidance on Pollution Prevention and Control of Smoke from Coal-Burning   《关于防治煤烟型污染技术政策的规定》(1984)

  the Ambient Air Quality Standards   《环境空气质量标准》(GB 3095—2012)

1.13 Water Pollution

  the Standards of Drinking Water Quality   《饮用水水质标准》(1956)

  the Regulations of Domestic Drinking Water Hygiene   《生活饮用水卫生规程》(1959)

  the Standards of Domestic Drinking Water Hygiene (on Trial)   《生活饮用水卫生标准》（试行）(1976)

  the Implementation Provisions for the Law of the People's Republic of China on Water Pollution Prevention and Control   《水污染防治法细则》(1989)

  the Environmental Quality Standards for Surface Water   《地表水环境质量标准》(1983)

  the Comprehensive Wastewater Emission Standards   《污水综合排放标准》(1988)

  the Regulations of Management for Pollution Prevention and Control of the Protecting Zones of Drinking Water Sources   《饮用水水源保护区污染防治管理规定》(1989)

  the Standards of Ground Water Quality   《地下水水质标准》(1993)

  the Emission Standards of Water Pollutants from Medical Institutions   《医疗机构废水污染排放标准》(2005)

  the Emission Standards of Water Pollutants from Saponin Industry   《皂素工业水污染物排放标准》(2006)

the Emission Standards of Pollutants from Coal Industry	《煤炭工业污染物排放标准》(2006)
the Ordinances of Urban Drainage and Sewage Treatment	《城镇排水与污水处理条例》(2013)
the Interim Ordinances of Water Pollution Prevention and Control for Huaihe River Watershed	《淮河流域水污染防治暂行条例》(1995)
the Measures of Management for Supervision on Effluent Outlets to Rivers	《入河排污口监督管理办法》(2004)
the Ordinances of Management for Water Drawing Permit and Water Resource Fee Collection	《取水许可和水资源费征收管理条例》(2006)
the Ordinances of Management for Taihu Watershed	《太湖流域管理条例》(2011)
the Action Plan for Water Pollution Prevention and Control	《水污染防治行动计划》(2015)
the Bulletin of the First National Census for Water	《第一次全国水利普查公报》(2013)
the Report Concerning the Status Quo of Water Pollution at the Guanting Reservoir and Recommended Solutions	国家计委、国家建委《关于官厅水库污染情况和解决意见的报告》(1972)
the Ordinances of Management for Water Supply and Utilizations of South-to-North Water Transfer Project	《南水北调工程供用水管理条例》(2014)
the Interim Measures of Management for Water Pollutants Emission Permit	《水污染物排放许可证管理暂行办法》(1988)
the Regulations on the Implementation of the Law of the People's Republic of China on Water Pollution Prevention and Control	《中华人民共和国水污染防治法实施细则》(1989)
the Construction Plan for the Urban Wastewater Treatment and Recycling Facilities during the 13th FYP	《"十三五"全国城镇污水处理及再生利用设施建设规划》(2016)

1.14 Land Resources and Soil Protection

the 2017 Statistical Bulletin of Chinese Land, Mineral and Marine Resources	《2017中国土地矿产海洋资源统计公报》(2018)
the Bulletin of the Fifth National Desertification and Sandification Land Monitoring	《第五次全国荒漠化和沙化土地监测公报》(2015)

the Bulletin of the First Nationwide Survey on Soil Contamination Status Quo　　《第一次全国土壤污染状况调查公报》(2014)

the Action Plan for Soil Pollution Prevention and Control　　《土壤污染防治行动计划》(2016)

the General Planning for the Comprehensive Survey of Nationwide Soil Pollution Status Quo　　《全国土壤污染状况详查总体方案》(2016)

the Notice of Several Policy Measures Regarding the Enhancement of Cultivated Land Protection and the Promotion of Economic Development　　《关于加强耕地保护促进经济发展若干政策措施的通知》(2000)

the Ordinances for the Resources Taxes (Draft)　　《资源税条例（草案）》(1984)

the Interim Ordinances of Farmland Encroachment Tax　　《耕地占用税暂行条例》(1987)

the Decision Concerning the Deepening Reform of Stringent Land Management　　《关于深化改革严格土地管理的决定》(2004)

the Interim Ordinances of Urban Land Utilization Tax　　《城镇土地使用税暂行条例》(1988)

the Recent Work Arrangements for Soil Environment Protection and Comprehensive Governance　　《近期土壤环境保护和综合治理工作安排》(2013)

the Action Plan for Zero Growth of Chemical Fertilizers Consumption by 2020　　《到2020年化肥使用量零增长行动方案》(2015)

the Action Plan for Zero Growth of Pesticides Consumption by 2020　　《到2020年农药使用量零增长行动方案》(2015)

the Guidelines for Environmental Planning　　《环境规划指南》(1994)

1.15　Terminology & Abbreviation

Centralization of Pollution Control	集中治理
Circular economy	循环经济
Cleaner production	清洁生产
Deadline for Pollution Governance	限期治理
Ecological civilization construction, economic construction, political construction, cultural construction and social construction shall be thoroughly integrated to build a sustainable and beautiful China (the "Five-in-One" Strategy)	五位一体建设

Emission Fees for Noncompliance	超标排污费
Emission Permit	排污许可证
Environmental economic policy	环境经济政策
Environmental Impact Assessment (EIA)	环境影响评价
EIA for plan (PEIA)	规划环境影响评价
EIA for project (Project EIA)	项目环境影响评价
Environmental monitoring and emergency management system	环境监测和应急管理
Environmental planning	环境规划
Environmental protection plan	环境保护规划
Information disclosure	信息公开
Liability System for Environmental Protection Objectives	环境保护目标责任制
One-Control-and-Two-Attainment	一控双达标
Pollution prevention and control	污染防治
Public participation	公众参与
Quantitative Performance Evaluation on Comprehensive Environmental Governance for Urban Environment	城市环境综合整治定量考核
Regional EIA (REIA)	区域环境影响评价
Strategic environmental assessment (SEA)	战略环境评价
the Central Committee of the Communist Party of China (The CCCPC)	中国共产党中央委员会
the Communist Party of China (The CPC)	中国共产党
the Environmental Protection Commission of the State Council (The EPC)	国务院环境保护委员会
the International Finance Corporation (The IFC)	国际金融公司
the International Financial Organizations (the IFO)	国际金融组织
the Leading Group of Environmental Protection (The LGEP)	环境保护领导小组
the Liability System of Environmental Protection Objectives	环境保护目标责任制
the Ministry of Ecology and Environment (The MEE)	生态环境部

the Ministry of Environmental Protection (The MEP)	环境保护部
the Ministry of Water Resources (The MWR)	水利部
the National Environmental Protection Administration (The NEPA)	国家环境保护局
the National People's Congress (The NPC)	全国人民代表大会
the Reform and Opening-Up Policy	改革开放政策
the State Construction Commission (The SCC)	国家建设委员会
the State Economic and Trade Commission (The SETC)	国家经济贸易委员会
the State Environmental Protection Administration (The SEPA)	国家环境保护总局
the Statistical Bulletin of National Economic and Social Development	国民经济和社会发展统计公报
the State Planning Commission (The SPC)	国家计划委员会
the United Nations Environment Programme (the UNEP)	联合国环境署
three synchronizations	三同时
Total Emission Cap	污染物排放总量控制
Trans-Century Green Project Plan	跨世纪绿色工程规划
Treatment happens after or as the pollution goes	边污染边治理，先污染后治理